AF692579

The Creative Mathematics Teacher's Book of Lists

The Creative Mathematics Teacher's Book of Lists

By

Peter Appelbaum

BRILL

LEIDEN | BOSTON

Cover illustration: Michael Georgeson

All chapters in this book have undergone peer review.

The Library of Congress Cataloging-in-Publication Data is available online at https://catalog.loc.gov

Typeface for the Latin, Greek, and Cyrillic scripts: "Brill". See and download: brill.com/brill-typeface.

ISBN 978-90-04-68845-2 (paperback)
ISBN 978-90-04-68844-5 (hardback)
ISBN 978-90-04-68846-9 (e-book)
DOI 10.1163/9789004688469

Copyright 2024 by Peter Appelbaum. Published by Koninklijke Brill BV, Leiden, The Netherlands.
Koninklijke Brill BV incorporates the imprints Brill, Brill Nijhoff, Brill Schöningh, Brill Fink, Brill mentis, Brill Wageningen Academic, Vandenhoeck & Ruprecht, Böhlau and V&R unipress.
Koninklijke Brill BV reserves the right to protect this publication against unauthorized use. Requests for re-use and/or translations must be addressed to Koninklijke Brill BV via brill.com or copyright.com.

This book is printed on acid-free paper and produced in a sustainable manner.

Advance Praise for *The Creative Mathematics Teacher's Book of Lists*

"This is definitely a different kind of book! It presents an innovative and thought provoking approach that will engage readers with an amazing array of ideas in mathematics education. I can see this book being used in many settings, from the individual reader to groups of teachers as part of a book study group to university courses. The different chapters invite readers to reflect and to pursue the ideas presented even further. It is a collection of appetizers that can be a meal in themselves but also open the door to delicious entrees."
– Marta Civil, Professor of Mathematics Education, Roy F. Graesser Chair, and University Distinguished Outreach Professor, The University of Arizona

"This groundbreaking book reveals the transformative power of exploring mathematics in the most unexpected of settings to change lives and reshape destinies. By cultivating inclusive spaces where every voice is heard and valued, it challenges and redefines traditional educational boundaries. A compelling read for anyone passionate about the role of mathematics in engaging and enriching the human experience."
– Lou Matthews, Global Mathematics Creative and Founder of *InspireMath*; Director of Mathematics and Science at Urban Teachers, Johns Hopkins University

"As a writer of mathematics books – and now a music one – I am still constantly on the lookout for math books that stretch my own thinking. Thankfully, there are a few. However, it is rare for me to come across such a book that I also wish I could have written. Earlier this year, such a book came into my orbit – *The Creative Math Teacher's Book of Lists*, by Peter Appelbaum. With such a title, one would assume that it is an easy book to describe. It is not. That was my first attraction to it, and subsequently why I fell deeply in love with it. This book is a meandering lifesource for boundless curiosity and wisdom for mathematics, reaffirming in no uncertain language that mathematics' allure and illuminations are available to all of us. There are so many ways to read, use, and share this book, reflecting and refracting all the power, magic, and humanness of the world of mathematics. Easily my favorite math book I have come across in years – I just wish I wrote it!"
– Sunil Singh, International Speaker & Math Storyteller, Author of *Pi of Life: The Hidden Happiness of Mathematics*, and *Chasing Rabbits: A Curious Guide to a Lifetime of Mathematical Wellness*

To
Charoula Stathopoulou,
the most creative and caring mathematics educator I know

Contents

Acknowledgment

Michael Georgeson is a visual communicator and creative problem solver specializing in animation, graphic design, and illustration. He is a graduate of the Rhode Island School of Design and holds a certificate in design from the Massachusetts College of Art and Design. A proud transgender man, his clients have included Yahoo (including the 2023 redesign of their Pride Month Logo), MIT, PJ Library (including animation for one of their most successful videos, *The Passover Story in Ten Scenes*), Blue Horizon Counseling, Princess Problems (album art), and Rogue Burlesque. He is passionate about using design to foster greater equity, accessibility, serenity, and justice in the world, and lives in Philadelphia with a dog and a cat.

I thank Vignesh and the typesetters at Brill for all of their herculean efforts in dealing with my never-ending format requests and updates!

About the Author

Peter Appelbaum
is Professor of Mathematics Education, and Director of Education Studies and Art Education Programs at Arcadia University in suburban Philadelphia, USA. He has worked for over 40 years in the preparation and professional development of teachers and as a musician/composer. An innovator with new and unusual ways to use and learn mathematics as the art that creates communities characterized by joy, a can-do attitude, and the courage to act on their convictions, he is the author, among others, of *Embracing Mathematics: On Becoming a Teacher and Changing with Mathematics* (Routledge), and *Popular Culture, Educational Discourse, and Mathematics* (SUNY Press).

Appelbaum identifies his undergraduate major in ethnomusicology and world music as solidifying his culturally embedded orientation to education through local conceptions of time and space in our postcolonial world. His subsequent doctoral studies in mathematics and in the social and philosophical foundations of education helped him to think deeply about mathematical ways of being and to embrace the pursuit of equity and diversity in his work. Most recently he is involved with queer and indigenous conceptions of mathematics education, reforms of prison education world-wide, and the promotion of youth leadership for mathematics and environmental post-human education through the Youth Mathematician Laureate Project.

INTRODUCTION

How to Read This Book – A List of Reading Approaches

- Teacher List Circles
- Leave It Lying Around
- Choose-Your-Own Adventure
- Re-gift the Gift That Keeps on Giving
- University Syllabus

Books of lists can be useful resources. Most of them are lists of things that are hard to remember. Fun facts. Weird trivia. Formulas and conversion rules. "Best of …" lists are fun, as are "The worst …" lists, because you can interact with them, agreeing or disagreeing with what is on the list, what the author forgot to include, even the order of the items on the list.

This book is not quite like those.

© PETER APPELBAUM, 2024 | DOI:10.1163/9789004688469_001

In your hands is a compendium of provocative lists. They are short, so you can remember them. However, they are not trivia or facts. They are always practical in the end, but some start out as very theoretical. Each list has bite-sized morsels of my personal take on teaching and learning mathematics. I believe they are *important*, in the sense that they have made a huge difference in my own teaching, and they have provoked powerful conversations with other teachers and with students with whom I have shared them over the past twenty or thirty years. I like to think of each list as a kind of "gift" – not like a scarf or a new bracelet; more like a new idea or toy that you can play with as you like, try to find new uses for, invent new versions of. Some are more like recipes that you can start with, and then tweak the ingredients or the length of time it takes to cook them; like a delicious meal, you can share them with others, and I hope you will!

What I have in mind is that you can decide for yourself how to best use these lists. One at a time? All of them at once? Here are some suggestions:

Teacher List Circles

My first suggestion is to invite a few teacher friends to read the book with you. As described in the section of this book called "Daniels-Type Literature Circle Roles", an excellent way to organize a book discussion group is for the group itself to decide how they want to read the book. Start at the beginning and get together after each section to discuss it? This might work well. Maybe one list per month is a good tempo. Each group member could experiment, trying to make some of the ideas in the selected list "work" in their classroom, modifying their practice each week or every other day. Then the group could get together over their beverage of choice (cafes and bars are good locations!), share stories and student work samples, and talk about new strategies for the next month or year, based on the shared experience. Maybe take a month off before starting a new list?

Do you need to go in the order of this book? No. Some teacher circles will rotate among members who will look through the book and select the next list. Others will read the entire book before their first meeting, and discuss which list they want to focus on for the week, month, or year. Still others will critique the list or the accompanying commentaries, like we all might do with those "best of ..." and "worst of ..." lists, and write to me with strong opinions – ooh! I will love those emails (Peter Appelbaum's email is appelbap@arcadia.edu) or letters! I will meet, virtually or in person, if we can organize it, to thank you, and to work on the revision of that list for the next edition!

Leave It Lying Around

Another way to use this book is to leave it in a place where you will see it a lot, pick it up when you can, open it to a random page, and think about that section. Perhaps it will be on your desk in your classroom – or in your bathroom at home, on a coffee table, or in the teachers' lounge library. Read one list and its commentary. Just spend a few minutes thinking:
- How do I already do this?
- How would I have written this section differently?
- What is missing? Why do I think Peter left that out?
- Who do I imagine might be someone that I could talk about this with? When would be a good time?
- What can I do with this in my classroom? – To increase student engagement? To make my teaching more interesting or fun? To better communicate with my students?

If you leave this in the teachers' lounge, why not play a game with your colleagues: Someone opens it up to a random page, and you talk about that list. Then, someone else turns the number of pages equal to that days' date, and you all look at *that* list – in the middle of its commentary if you are on that page (if you are near the end of the book, you may have to use a sort of wrap-around counting to continue from the beginning of the book ...).

Choose Your Own Adventure

With this approach, you select a list as a serious focus for an extended period of time and pursue its ideas as an action-research project. You not only read the list and its commentary, but the "Future Reading" suggestions as well. In response, you think of 1–3 specific things you will introduce into your teaching in order to integrate the ideas into your classroom practice. Each week, you look back on how things went, and in a journal or other tool for reflecting, you record what worked well, what didn't, what your new questions are, and what you will do in the following week to pursue these new questions. Once a month, you look back over the list and your notes on the "Further Reading" items, and think about how you can take a fresh approach to this adventure project.

After several months of this, you get together with another teacher to tell them about what you are doing, and to get their feedback. This meeting will be useful because it will help you express your personal reflections in a way that someone else can understand. Go into those "Further Reading" materials, and find 1–3 references in them that you can consult as additional resources for your adventure project. Use these new resources to inspire new experiments for your classroom. Include your students in this adventure, since they will likely have fantastic suggestions.

Re-gift the Gift That Keeps on Giving

Give a copy of this book to a friend, family member, or an administrator in your school. Mark a specific list as one you tell them you thought they might find interesting, surprising, outlandish, or the solution to their utmost concern. See if they come back to you in a week or more. If not, you might try lightly asking them what they thought of the book – have they looked at it at all? Why or why not?

University Syllabus

Last but not least, this book could be a refreshing textbook for a course about learning and assessment in mathematics at any level! If so, I would assign one or two lists per class meeting as the introduction to the topic. Students should read the "Further Reading" suggestions along with the instructor's own additional materials for elaboration upon the basic ideas. A nice assignment for students is to create their own list on a topic or set of issues not included in this book, along with a brief commentary and suggested "Further Reading" resources. In this way, the course stays fresh and relevant to current issues and trends in mathematics education; the book provides a canonical perspective on the field from the past 30 or so years, and the students are building their own living textbook by adding their own sections for each other.

One More List of Advice

I wish you fun surprises as you decide how you want to read this book! Stay in touch! As a further preview of what's in store, here are some thoughts:

- As in the section on "A Short List of Mathematical Games You Might Have Overlooked", where I suggest any game can be a template for a mathematics game, each section here is a template for your own lists. Just like the students in the university course idea, any reader should think of a list that deserves to be added to those in this book. Create the list. Note that the best lists are brief, so work on finding a way to shorten your initial list by combining ideas into more generalizable categories (I did this, and it was a challenge! – it's also a kind of way of working like a mathematician!). Add some explanatory commentary, and find a few suggested resources to consult for further provocation. Share your list using any of the suggestions in the above list of "How to Read This Book".

- You can create an inspiring "List Poem" by flipping through the book and randomly selecting one sentence from each page to which you flip. Organize your list so that it makes sense to you. Edit the phrases in your list poem to your liking. Post it somewhere that will be visible and use it as an affirmation: every time you see it, read it, and describe to yourself how you are enacting the core meaning of your poem on a daily, weekly or monthly basis. You might give a copy of the poem to another teacher as a silly (or serious?) "gift". Trade poems. Put together a chapbook of the poems in your Teacher List Circle.
- Some of the people mentioned in this book are alive and well, and very active as teachers and scholars! Google them. Look for their names in online research databases and find newer things that they are writing about. Read some of their more recent work! Also, I know almost all of them personally, and believe that they would be tickled pink to receive an email message from a fan. They are very likely to reply or continue a virtual conversation with you! Try it.
- A podcast or YouTube channel where you or your class take on one list at a time and "review it and its applications" would be a swell idea. If you are critical enough, so that you are adequately building on my initial ideas, I will be so delighted I will beg of you to be a guest on your show!
- Above all, have a good time with these lists. It may not always be clear to every reader, but I mostly expect you to be laughing a lot. Enjoy your time with this book!

Happy Reading and Thinking!
Best wishes,
Peter Appelbaum
Summer 2024

CHAPTER 1

Mathematical Activity Lists

- Alan Bishop's Six Mathematical Activities
- Tamsin Meaney's List of Bishop's Six Activities
- David Kirshner's Three Types of Activity

Alan Bishop's Six Mathematical Activities

Bishop (1988) identifies six "universal" activities that can be characterized as mathematical. In addition, he also specifies for each activity some "organizing concepts" that should provide a "knowledge frame:"
- Counting
- Locating
- Measuring
- Designing
- Playing
- Explaining

© PETER APPELBAUM, 2024 | DOI:10.1163/9789004688469_002

Alan Bishop, Professor Emeritus at Monash University in Australia, received the Felix Klein Medal from the International Commission on Mathematical Instructions (ICMI) in 2015, an honor bestowed only on the most distinguished and impactful researchers and scholars in mathematics education. The award recognized his more than forty-five years (now well over 50!) of sustained, consistent, and outstanding lifetime achievements in mathematics education research and scholarly development.

Alan's early research on spatial abilities and visualization took on an entirely new framing during his 1977 sabbatical leave in Papua New Guinea, during which he began a lifelong quest to understand the processes of mathematical enculturation, and how they take on different forms and meaning in different countries and cultures. His landmark 1988 book, *Mathematical Enculturation: A Cultural Perspective on Mathematics Education*, was revolutionary and introduced a new conception of mathematics – the notion of mathematics as a cultural product, embodying and structuring cultural values. He sparked further evolution of this orientation by co-organizing a special day-long event during the 1988 Sixth International Congress on Mathematical Education devoted to "Mathematics, Education, and Society", which led eventually to successive conferences on the political and social dimensions of mathematics education. Every mathematics teacher should know of Alan Bishop and his significant work in bringing the political, social, and cultural dimensions of mathematics education to the attention of the field.

The basic take-away from this list, "Alan Bishop's Six Mathematical Activities", is that we should shift our perspective on what mathematics is, de-emphasizing mathematics as a collection of things you do, to mathematics as a way of knowing and being. Mathematics is a human activity that all students have been slowly absorbing as part of their everyday life. It is not so much arcane and special symbols and procedures and algorithms – instead, those symbols and procedures and algorithms are helpful ways that some people have developed to communicate with each other about rather mundane and universal experiences. Children and young adults bring to school many rich and highly developed ways of making sense of their lives and their world. Mathematical ways of being are deeply embedded in the ways that all human beings think and act. School mathematics may look different on the surface, but this is the fun of teaching, to further enculturate our students into the power of the special forms of communication, and to help them to realize their dreams through this further refinement of what they already know as reality and common sense. If this strikes you as really weird, then, yay, welcome to the exciting world of anthropological approaches to mathematics teaching!

Counting

Quantifiers (each, some, many, none); Adjectival number names; Finger and body counting; Tallying; Numbers; Place value; Zero; Base 10; Operations on numbers; Combinatorics; Accuracy; Approximation; Errors; Fractions; Decimals; Positive, Negatives; Infinitely large, small; Limit; Number patterns; Powers; Number relationships; Arrow diagrams; Algebraic representation; Events; Probabilities; Frequency representations.

Locating

Prepositions; Route descriptions; Environmental locations; N.S.E.W. Compass bearings; Up/down; Left/right; Forwards/Backwards; Journeys (distance); Straight and Curved lines; Angle as turning Rotations; Systems of location: Polar coordinates, 2D/3D coordinates, Mapping; Latitude/Longitude; Loci; Linkages; Circle; Ellipse; Vector; Spiral.

Measuring

Comparative quantifiers (faster, thinner); Ordering; Qualities; Development of units (heavy – heaviest – weight); Accuracy of units; Estimation; Length; Area; Volume; Time; Temperature; Weight; Conventional units; Standard units; System of units (metric); Money; Compound units.

Designing

Design; Abstraction; Shape; Form; Aesthetics; Objects compared by properties of form; Large, small; Similarity; Congruence; Properties of shapes; Common geometric shapes, figures and solids; Nets; Surfaces; Tessellations; Symmetry; Proportion; Ratio; Scale-model Enlargements; Rigidity of shapes.

Playing

Games; Fun; Puzzles; Paradoxes; Modelling; Imagined reality; Rule-bound activity; Hypothetical reasoning; Procedures; Plans Strategies; Cooperative games; Competitive games; Solitaire games; Chance, prediction.

Explaining

Similarities; Classifications; Conventions; Hierarchical classifying of objects; Story explanation; logical connectives; Linguistic explanations: Logical arguments, Proofs; Symbolic explanations: Graphs, Diagrams, Charts, Matrices; Mathematical modelling; Criteria: internal validity, external generalizability.

How do we facilitate our students in their use of mathematical ways of knowing and being to make sense of their world, to make decisions and choices in their lives, to appreciate ethical dilemmas and resolutions, and so on? We begin by understanding that they already do this, even if it looks very different to us! Early in a school year or in the initial lessons of a program, unit, etc., we need to set the learners on an adventure that we can observe. During an open-ended investigation or a classroom conversation, we take note: How are they "counting", in the broadest sense of this list? Locating? Measuring? Designing? Playing? Explaining? Recognize that Bishop's use of these terms is far more expansive than our typical and simplistic articulations of them. Counting is not just saying number names one after the other, but all of the many extensions of that early childhood experience. Our students are likely demonstrating some or all of the many things in what he groups together as "counting". Similarly, "playing" is not merely recreational passing of time but includes rule-governed structured experiences and the abstraction of patterns and strategies that can be employed within such experiences.

This book does not lay out all the implications of its lists. Instead, it invites you to try out the lists for reflecting on your practice. How can you plan a unit, a week of lessons, or a year-long trajectory in ways that facilitate each of the six categories of mathematical activity? How can your students take advantage of these ways of being and knowing to learn more about themselves and others? How can they build upon the informal forms of these activities that they have been experiencing their entire lives, in order to continue their enculturation into the world of conventions and expectations that their school mathematics culture demands of them?

Finally, it has become increasingly apparent since the 1980s that mathematical enculturation is accompanied at all times by what anthropologists call "acculturation". ENculturation refers to all of the life experiences in and out of school through which young people are enabled to become more sophisticated, grown-up members of that culture over time. ACculturation highlights the power relations between those who are more or less experienced or sophisticated. Acculturation also recognizes that there are some people whose identity is more closely affiliated with the dominant, school mathematical culture; these people have advantages in these power relations. "Learning" in our classrooms might be suitably understood as enculturation, acculturation, or both.

Further Reading

Appelbaum, P., & Stathopoulou, C. (2015). Critical issues in culture and mathematics learning. In L. English & D. Kirshner (Eds.), *Handbook of international research in mathematics education* (pp. 744–798). Taylor and Francis.

Tamsin Meaney's List of Bishop's Six Activities

- Playing
- Explaining
- Measuring
- Designing
- Counting
- Locating

Tamsin Meaney is a professor of mathematics education at the Western Norway University of the Applied Sciences, in the Department of Language, Literature, Mathematics and Interpreting. She has been involved in teacher education programs and supervision of post-graduate students in New Zealand, Australia, Sweden and Norway. A highly productive mathematics education researcher with more than 100 peer reviewed publications, she specializes in Indigenous education, early childhood education, and teacher professional development. Currently the research leader of a large Norwegian Research Council funded project on "Learning about teaching argumentation for critical mathematics education in multilingual classrooms", she also runs a project as part of the Kindergarten Knowledge (Kindknow) Centre for Systemic Research on Diversity and Sustainable Futures. A popular keynote presenter world-wide, this list comes from a 2016 presentation she gave as part of The Early Childhood Seminar Series at the University of Auckland Faculty of Education in New Zealand.

This list – ordered a bit differently from Bishop's to highlight the activities she prioritizes – might be more easily implemented in your teaching than the original one from Alan Bishop. An important take-away from Tamsin Meaney's presentation, however, which she demonstrates through careful study of very young children in Swedish pre-schools, is that profound mathematical concepts are developed, much to our surprise, prior to having the language to express them. In her presentation, she focuses on Bishop's "Locating", and shows how young children's informal, autonomous navigation in space helps them to construct ideas about position, up, down, around, and through, in relations between where they are located and where they want to go – for example, in trying to get down from up high on a playground structure. Adults play an important part in the development of these concepts by accompanying the children during their experiences, facilitating the children's own desires to move and reposition themselves. Enculturation formats the possibilities for their life-world experience in very concrete ways, such as a grid-like shape to a playground structure, which presents Cartesian space as the way to interact with the world. The young children ask adults for assistance, and it is in these moments that the Swedish early childhood educators provide a specific sort of experience: they don't solve the problem for the child, such as by picking them up and carrying them down from the structure, but instead provide a safety net of security, standing by or gently holding them, as they figure out where to place their feet or hands, how to jump, etc. We can take this realization about language following conceptual development seriously by providing extended time with exploration and non-structured investigations with our materials (manipulatives, graphical representations, charts and so on), before demanding that students use language or representation to communicate about what they are experiencing. The need to communicate can, for older children, move more quickly toward finding words and pictures to explain their thinking, but it turns out that the more time learners spend on their own, exploring, and then only later asking for help, the better; the "help" is more of a standing by and assisting in the learners' own solutions to their dilemmas. Later on, at another time, telling the story of what was experienced or accomplished, the need for language and representation presents itself. Note in the following descriptions how she simplifies the activity qualities in order to help you think about the engagement of the students.

Playing
Devising and engaging in games and pastimes, with more or less formalized rules that all players must abide by.

Explaining
Finding ways to account for the existence of phenomena, be they religious, animistic, or scientific.

Measuring
Quantifying qualities for the purposes of comparison and ordering, using objects or tokens as measuring devices with associated units or 'measure-words'.

Designing
Creating a shape or design for an object or for any part of one's spatial environment It may involve making the object, as a 'mental template', or symbolizing it in some conventionalized way.

Counting
The use of a systematic way to compare and order discrete phenomena. It may involve tallying, or using objects or string to record, or special number words or names.

Locating

Explaining one's spatial environment and conceptualizing and symbolizing that environment, with models, diagrams, drawings, words, or other means.

Tamsin Meaney's work with the Kindknow Project has led to interesting ideas about how we should design and use interactive digital media in our teaching. She focuses on social justice in the development of apps that promote children's talking about mathematical ideas both at home and in school. It is critical that learners talk in both their mother-tongue and in the language most used in school, in order to best support multilingual learners. Immigrant families often use digital technologies more than non-immigrant families, so they are an interesting audience for working with teachers, learners and apps. Children, teachers and families designing games together better ensures that the formats for mathematics learning are better than those imagined by external researchers or professional designers.

Further Reading

Farsani, D., Lange, T., & Meaney, T. (2022). Gestures, systemic functional linguistics and mathematics education. *Mind, Culture, and Activity*, 29(1), 75–95.

Meaney, T., Helenius, O., Johansson, M., Lange, T., & Wernberg, A. (Eds.). (2016). *Mathematics education in the early years: Results from the POEM2 conference*. Springer.

Meaney, T., Trinick, T., & Fairhall, U. (2012). *Collaborating to meet language challenges in indigenous mathematics classrooms*. Springer.

Tamsin Meaney's paraphrase of Alan Bishop can be found in her presentation, "Locating learning of toddlers in the individual/society and mind/body divides", The University of Auckland, October 2016. https://www.youtube.com/watch?v=GfxmKZv9Kv8

David Kirshner's Three Types of Activity

- Exercises
- Puzzles
- Probes

David Kirshner (2000), Professor and Director of the Cain Center for STEM Literacy at Louisiana State University in the United States, noticed that teachers and students do not always recognize or agree on the nature or purpose of activity in the classroom. Teachers themselves aren't always clear on why they are asking students to participate in certain activities! Sometimes, we want students to practice skills; we ask them to do exercises in order to increase automaticity and to help them get comfortable with certain fact recall and procedures so that they can use them without too much thinking, that is, to focus on more important things. At other times, we pose a question for students because it is fun to think about, or because it demonstrates a key connection; these are puzzles. Much of the time is spent in yet another way, on investigations, explorations, and open-ended inquiry through what Kirshner calls "probes".

The thing is, we might think we are asking a student to explore an open-ended situation, yet the student approaches it like a puzzle, feels like they have no idea what the trick is, and stops thinking in order to wait for the answer or solution to be explained. They somehow missed the cues that would telegraph that they should just try some problem-solving strategies, or list attributes of the situation in order to pose their own problems. Or, maybe we want students to practice a skill with a collection of routine problems, and some of our students instead interpret this as a collection of similar problems that they should compare, contrast, categorize, and critique. Those students might learn fascinating stuff from turning the exercises into probes, but they are not practicing skills toward automaticity, and will not be able to complete the simple practice session within the allotted time.

There are two big takeaways from this awareness: First, we need to help our students recognize what sort of approach is expected of them when we introduce an activity. They need cues to know whether they should be practicing a skill, looking for a clever connection, or exploring an open-ended situation. One way to do this is simply to be very clear about what you are expecting, and to say so. "I have a puzzle for you. Does anyone see the trick?" or, "Here is a collection of exercises for you to use as you practice applying the binomial theorem to polynomials". "OK, so we have some first ideas and reactions ... now, for the next 20 minutes, use the time for exploring one of those ideas as an investigation with no right answers. We will gather together afterwards to share what we found, what our new questions are, and what help we request from others".

Second, and more important perhaps, is that each of these types of activity have cultural strengths that we can exploit, and weaknesses to avoid.

Exercises

... can create confidence and feelings of mastery. Automaticity and quick recall free up our minds for more interesting work. Unfortunately, they are also usually boring, since they are repetitive. They can become tedious. It's best if we can keep in mind why we are doing these exercises. A sports team will gladly spend an entire practice session on one drill if they have been shown video of previous games where they were weaker on their left side, and if they are told that these drills will help them become more agile on their left. A musician will practice an arpeggio carefully, slowly, and at faster and faster speeds, with as many different articulations as they can come up with, if they want to be able to play that arpeggio as a small piece of a musical phrase in the next week. Similarly, students might happily drill each other on math facts, or practice division of fractions, if they have been slowed down by these things in their ongoing investigations.

Exercises are not so good for informal assessment. Students either get the answers or they don't. Do they need more practice, or is there something you need to know about how they are thinking? You can't see inside their thinking through exercises. Nevertheless, students who can successfully complete a series of exercises in a short amount of time can achieve a strong sense of satisfaction, a sense of self-efficacy, and a clear demonstration of having mastered some material.

Puzzles

... are just that: they have a solution, and there is usually a trick to solving it. It seems you either know it or you don't. The solution is like finding just the right puzzle piece in a box of 1000 pieces that belongs in just the right place in a jigsaw puzzle. When you find it, you are very satisfied! When you don't, you are frustrated. Similarly, in a classroom, a mathematical puzzle presented by a teacher or one of the students has this quality of "you either get it or you don't". It runs the risk of shutting students down, as they tell themselves, "I don't know. I am not a math person". On the other hand, if the puzzle is at just the right level of challenge, someone in the room is probably going to see the trick to solving it. The key is that, once the secret is revealed, pretty much everyone does get it. What this does is create a feeling of belonging. "We get it, because we are in this class". Even if it has to be the teacher who shares the secret after students are given time to try, if almost everyone understands the solution, it can have this impact.

Puzzles are not so much a form of assessment on mathematics objectives, but they do provide some informal assessment of more amorphous aspects of the classroom. Are the students interested in the puzzle, or do they shut down? When a puzzle is pitched at the right level, meaning almost everyone can understand the solution even if they couldn't think of it themselves, are they excited and do they share it with others outside of their class? These are indications of a positive relationship with mathematics upon which they can build more conceptual understanding and further develop skill fluency. Puzzles also indicate levels of persistence. Some learners enjoy the challenge of a puzzle, because they anticipate the eventual pleasure of solving it. Teachers can also observe persistence, and support students in applying it. After all, the solution is only a small part of mathematical joy. We usually discover far more interesting things along the way, whether we solve the puzzle or not!

Probes

... are good for encouraging creativity, and for apprenticeship in the ways of working as a mathematician. They are not best for practicing skills or feeling like you are part of a special group. They are incredibly vital in the ongoing development of mathematical dispositions, and for the appreciation of mathematics as a way of being in the world. They are the essence of mathematics expanded into ongoing activity, they build community interactions, and they promote deep thinking. They might also be highly frustrating for students who expect to be told what to do. Such students might just sit back and wait, not knowing how to be in a room that does not require practicing exercises or appreciating puzzles.

Probes are excellent opportunities for a teacher to move about the room and carry out observations and other forms of informal assessment. What are students doing? Sharing ideas? Drawing diagrams? What concepts are they readily applying without a prompt? Which relevant skills are they *not* using when they should be? What student work could you request that a student share with the rest of the class later on? While there is no guarantee that any one skill or concept will be practiced during this sort of activity, there are many more opportunities to learn about how the students are thinking and making sense of the mathematics, how persistent they can be, how successful they can be working independently, and so on.

Further Reading

Kirshner, D. (2000). Exercises, probes, puzzles: A crossdisciplinary typology of school mathematics problems. *Journal of Curriculum Theorizing, 16*(2), 9–36.

Sullivan, P., Aulert, A., Lehmann, A., Hislop, B., Shepherd, O., & Stubbs, A. (2013). Classroom culture, challenging mathematical tasks and student persistence. In V. Steinle, L. Ball, & C. Bardini (Eds.), *Mathematics education: Yesterday, today and tomorrow* (Proceedings of the 36th annual conference of the Mathematics Education Research Group of Australasia). MERGA. https://files.eric.ed.gov/fulltext/ED573047.pdf

CHAPTER 2

Assessment Lists

- Alice Pitt Assessment Categories
- Susan Ohanian's Ten Foundational Questions
- Jo Boaler's Five Elements of Effective Struggle

Alice Pitt Assessment Categories

- Youth Leadership
- Voice
- Participation

There are so many things we might focus on in assessment, but we all too often obsessively look at student performance of skills and application of conceptual understanding. Sometimes, it can be powerful to shift our focus onto other, perhaps more important aspects of the classroom culture, in order to make even greater strides in our usual skill and concept learning goals. Alice Pitt, a

© PETER APPELBAUM, 2024 | DOI:10.1163/9789004688469_003

philosopher of education and currently the Interim Vice President for Equity, People and Culture at York University in Canada, offers three areas to look at as a facilitator of learning (Pitt, 2003). Learners are placed in positions of leadership. Lessons and other activities are planned to increase student voice and participation. Ongoing reflection seeks to include those not yet leading and participating, as well as to expand the kinds of leadership and participation that are possible.

Assessment ideas include: (1) Designing activities so that each student, over time, once a day, week, month, unit, etc., facilitates the discussion or decides the topic, or organizes a project. What forms do their leadership take? How are they reflecting on their satisfaction with their leadership style? What supports do they need to take on additional forms of leadership? You can support students by keeping records that you can share with the students; by holding mini conferences to discuss their journal entries about their leadership experiences; and by facilitating whole class discussions that identify effective strategies for discussion leaders, group leaders, and topic selectors. (2) Ensuring student contributions push the conversation into new directions, so that they recognize that their participation changes the direction of lessons – a question or sharing of a confusion is a greater contribution than a correct answer, for example; by keeping track of who is talking or sharing work, and making sure that all students' voices are part of large and small group conversations; by requiring small groups to make sure that each member can explain what they are doing and why, walking around the room and testing each group – if a member cannot explain what is happening, demand that the group help that member to be ready to explain to you when you return in 5 or 10 minutes. (3) Planning regular organization of classroom activity to require active participation and engagement. Have students keep track for themselves of the ways they have participated and engaged in the classroom, and have routine weekly or monthly reflections on their satisfaction with their participation and engagement, including a plan with specific actions they can take before the next check-in.

Pitt writes eloquently about these categories of assessment in her book, *The Play of the Personal*. Something to keep in mind is how the classroom "is implicated in the very social relations it hopes to observe, analyze, and disrupt" (p. 53). These sorts of pedagogy that we are describing here have the power to disrupt students' prior self-understandings, and might be initially experienced as bizarre, inappropriate, or confusing. Such disruptions, writes Pitt, risk producing new and often debilitating forms of helplessness and isolation. Resistance to these forms of assessment, might in this respect, "represent a mode of interpretive negotiation of the social as it is lived in classrooms" (p. 53). So, preparing for this potential is part of the assessment plan design.

Further Reading

Pitt, A. (2003). *The play of the personal: Psychoanalytic narratives of feminist education.* Peter Lang.

Susan Ohanian's Ten Foundational Questions

1. Do my students see themselves as mathematicians?
2. Do my students see mathematics as covering a wide range of topics?
3. Are my students developing a flexible repertoire of problem-solving strategies?
4. Are my students able to communicate their problem-solving strategies to others? Can they talk and write about how they solve math problems?
5. Are my students able to assess themselves? Are they able to develop and use criteria to evaluate their performance?
6. Do my students engage in mathematical thinking without a specific assignment? For example, of they have "free time", do they choose math?
7. Are my students developing the attitudes of independent and self-motivated thinkers and problem solvers?
8. Do my students welcome challenges in math? Are they able to focus on math problems of increasing complexity for longer periods of time?
9. Do my students recognize the importance of math in the real world outside school?
10. Do my students use mathematics to solve problems outside math class? (Adapted from Ohanian, 1992)

Long-time teacher and prolific free-lance writer/journalist Susan Ohanian has published widely, ranging from the *Atlantic* and *Washington Monthly* to *Phi Delta Kappan* and *Education Week*.[1] Based on one of her many exhaustive research studies visiting many classrooms, she came up with the above list of ten foundational questions that every teacher should use to reflect on their teaching. These questions form the basis of a second version of assessment following Alice Pitt. For each question, what you need to do is come up with ways that you can collect information that will help you answer that question.

Do your students see themselves as mathematicians? What would it look like if they did? If we use Alan Bishop's mathematical enculturation to guide us, we would see students independently, without being cajoled, applying techniques of counting, locating, measuring, designing, playing and explaining, to carry out productive work. They would spontaneously suggest these approaches to thinking and living in the classroom and tell stories of doing these things in their everyday life without prompts. This means that your classroom must create opportunities for you to see if they do indeed do these things.

Do your students see mathematics as covering a wide range of topics, or would they instead respond to a visitor who asks them "what mathematics is" with the name of the current skill or concept being learned? What ways does your classroom facilitate students seeing the connections among seemingly disconnected math concepts and topics? The more you integrate this into your teaching and assessment, the more you will be able to see if your students are making progress in this area.

Are your students developing a flexible repertoire of problem-solving strategies? Are they able to communicate their problem-solving strategies to others? Can they talk and write about how they solve math problems? The more you expect students to identify strategies on their own, share them, compare them with other strategies, and experiment with strategies that others have used, the more you can assess this. An excellent expectation is that students can not only solve a problem on their own, but can explain their thinking well enough to another student so that the other student can understand it and use this different strategy themselves; similarly, they should be able to listen to an explanation of another student's thinking, ask enough clarifying questions, and be able to use this other strategy themselves. Then these students should describe what the strategies have in common, how they differ, and when or where each has its merits.

Are your students able to assess themselves? Are they able to develop and use criteria to evaluate their performance? The more you incorporate practice on this, including discussions of what successful self-assessment looks and feels like, how to cope with self-critical reflection, and so on, the better your students will be at such action. Do they engage in mathematical thinking without a specific assignment? For example, if they have "free time", do they choose mathematical activities? Do they incorporate mathematical ways of thinking and being within the context of other curricular areas, such as when they are reading literature or studying social studies? Do they recognize the importance of math in the real world outside school? Do students use mathematics to solve problems outside math class? They can demonstrate this if they are given opportunities to use mathematics to understand their world, take action in it, and explore new ideas with mathematics.

Are students developing the attitudes of independent and self-motivated thinkers and problem solvers? Do they welcome challenges in math? Are they able to focus on math problems of increasing complexity for longer periods of time? What does this look like? Perhaps we can assess this by seeing a reduction over time in the need for constant guidance, in fewer and fewer requests for assurance that they are correct. The more they autonomously initiate their own questions of each other, the more they are acting like a community of mathematicians.

Further Reading

Nielsen, K., & Ohanian, S. (2015). *Resisting reform: Reclaiming public education.* Information Age Publishing.

Ohanian, S. (1994). *Who's in charge? A teacher speaks her mind.* Heinemann.

Jo Boaler's Five Elements of Effective Struggle

- Teach about the Brain's Ability to Grow
- Model & Praise Mistakes as Opportunities for Brain Growth
- Provide Rich, Open-Ended Math Tasks
- Promote Deep Thinking: Remove an Emphasis on Speed
- Be Mindful of Your Own Relationship with Mathematics

Important questions to ask yourself are whether, when and how your students are struggling as they are learning in your classroom. This list provides five orientations to assessment that you can use to make sure that struggle is taking place, that it is productive, and that your students are learning to appreciate struggle as a valuable – and even in a certain sort of way enjoyable – feature of learning and doing mathematics. As Stanford University professor of mathematics education Jo Boaler notes in her book, *Limitless Mind: Learn, Lead and Live without Barriers,* if you aren't struggling, you aren't really learning. Nobody is born with a special brain. People who are highly successful in any set of skills or any realm of knowledge engage in a very particular kind of practice. People who achieve the most are always pushing at the edge of their understanding. Making mistakes, correcting them, and making more mistakes are all part of how they learn. High achievers are always struggling toward constantly revised, high targets. Whenever they become stuck, they try a different approach, so that they can keep going. As she also notes, there are no limits to what people can achieve. If people stop achieving, it is probably because they have given up, and not because they have hit a genuine roadblock in their pursuit of their high goals. Our trick as teachers is to help students reorient themselves to struggle. If they are struggling, we want them to know that this is the best feeling for their brain. When they are struggling, their brain is growing.

Margaret Smith and Mary Kay Stein (2018) similarly encourage teachers to pose purposeful questions, elicit and then turn around and use evidence of student thinking, and to facilitate engagement with and connecting among multiple mathematical representations to promote "productive struggle". Their description of classroom discussions includes ways that teachers can support students' interaction with mathematics without doing the thinking for them. The idea is to communicate through your actions the message that you truly believe your students are capable of making progress on their own. Each and every student, then, regardless of what they bring to any lesson, has access to the task and time to work without interference from a well-meaning teacher or "helpful" group members.

Five ways we can help students to embrace struggle begin with directly explaining the neuroscience research. We can show videos about the latest research, interview local researchers, and share biographies of mathematicians who struggled on their way to achievement. Jo Boaler has such videos freely available on her YouCubed website. We can share our own stories of struggling on the road to success, and model positive responses to our own mistakes, describing how they are opportunities for learning. Indeed, mistakes more than lucky success are far more helpful for understanding important ideas. Analyzing mistakes together in class can reveal very powerful connections. Studying mistakes that led to significant accomplishments provides a context for all of this.

The playbook lists in this book create a welcoming environment of open-ended tasks that make mistakes more useful than lucky guesses. They also make it clear that deep thinking is far more powerful than speed. Quick calculations and recall of memorized procedures are certainly valuable, because they allow us to focus on the new and important things at hand. Yet what they help with is making space for *slowing down* and thinking about things carefully. Mathematicians are some of the slowest thinkers around, because they have had to pay attention to things others overlook on their race to answers. They also need to keep their antennae out for new questions that others have missed, new opportunities to make connections and new solutions.

A certain kind of dismissal of the "easy way out" is crucial to the disposition we are seeking to nurture in our classrooms when we take this approach. If we reduce confusion in the name of "learning without even realizing it", we have lost the chance, as the German poet Rainer Maria Rilke once put it, to "live the questions" (Rilke, 1954). "The more fully we experience what is difficult", he wrote, "the more it pulls and drives us with its weight toward the center of life". What Rilke once advised is apt for the poet-mathematicians in our classrooms: "… have patience with everything", and "try to love the questions themselves as if they were locked rooms or books written in a very foreign language".

Philosopher of education Mordechai Gordon explains that Rilke wanted us "not only to accept our own doubts, confusions, and uncertainties as integral parts of who we are, but also to embrace them as conditions that are valuable in themselves, which need to be cherished and protected" (Gordon, 2007). With Rilke, we can encourage our students to appreciate our questions, doubts and confusions as valuable experiences, because facing them often leads to intellectual, emotional, and spiritual development. The locked rooms or books written in a "very foreign language", writes, Gordon, are metaphors for the mysterious and inexplicable nature of human existence. The point, he suggests, is to cherish and celebrate those perplexing or even incomprehensible moments in our lives at least as much or more than those that are clear and coherent.

Finally, we need to reflect on our own orientation to mathematics. How are we speeding along to get things done rather than slowing down to appreciate the mathematical scenery all around us? What new investigations might we have enjoyed if we had taken the time to think? Psychologist Carol Dweck (2007) reminds us, every time we make a mistake, we grow a synapse. Many of us were drawn to mathematics because we mistakenly believed in its promise of certainty and security. Rilke would want us to explore the emotional insecurity that underlies a quest for certainty in teaching, or certainty in the answers to our mathematical questions. Whenever we are doubting the apprenticeship of our students in productive struggle, it is a moment we can use to closely examine our own anxieties and fears that push us in the direction of controlling our classes, rather than falling back on a search for best techniques and perfect lesson plans.

Note

1 Susan Ohanian maintains a blog at https://susanohanian.net/, and a website, www.susanohanian.org, which holds tons of information on education issues.

Further Reading

Boaler, J. (2019). *Limitless Mind: Learn, lead and live without barriers*. HarperOne.

Boaler, J. (n.d.). *The importance of struggle*. https://www.youcubed.org/resources/the-importance-of-struggle/

Smith, M., & Stein, M. K. (2018). *5 practices for orchestrating productive mathematics discussions*. NCTM.

CHAPTER 3

Hannah Arendt Lists

- Hannah Arendt's Three Kinds of Activity
- Hannah Arendt's Four Strategies for Unlearning

Hannah Arendt's Three Kinds of Activity

- Work
- Labor
- Action

How do we understand the nature of acting as a mathematician in our classroom? According to philosopher and journalist Hannah Arendt (Arendt, 1958), it has to do with the use of the mathematics related to the relationships among the people involved. She described activity as taking on the character of "work" when the activity is intimately related to conditions of life. The same activity would be called "labor" if the effort and time were removed from those conditions of life and of being human, or, in contrast, "action", if people come

© PETER APPELBAUM, 2024 | DOI:10.1163/9789004688469_004

together in community through the activity. For example, if one makes a chair in order to sit on it, the making of the chair is work. If one makes a chair in a factory so that the corporation can sell the chair to others, then the making of the chair is labor. If groups of people make chairs so that they can sit together and talk, eat, sing, and so on, then the making of the chairs is action.

Work

If the mathematical activity is intimately connected to the conditions of life, we call it "work" (as opposed to purposeless passing of time).

Labor

If the mathematics is serving others, such as producing materials for assessment, then this is labor, not genuine "work".

Action

If the mathematics contributes to community-building, then it is "action" – the most valuable form of activity.

We can use this language to describe the activity in a mathematics classroom, or other learning contexts. How is the use of mathematics related to the relationships in this collection of people? For example, if a learner is practicing the application of an algorithm, then this learner is working. If the work is only serving others, such as when learners produce materials for assessment, then this is labor, and not genuine "work". If the work contributes to community building, then the activity is action, such as when a group of students prepare a presentation of what they have been exploring. Arendt's categories of analysis allow conversations about classroom community to connect with issues of power and knowledge in a broader social context, alongside more commonplace analysis of performance related to content objectives, the development of mathematical thinking, forms of argument, and empirical research. In this way, the analysis of mathematics classrooms and the forms of knowledge production contribute to the ongoing development of pedagogical methods and the design of classroom materials.

Analyzing mathematics teaching and learning in terms of work, labor and action might seem like an entirely new way to study, assess, or design mathematics curricula. However, there is a strand of mathematics education research that has used analytic categories in similar ways. For example, Kirshner (2002), whose Exercises, Puzzles and Probes were dicussed in Chapter 2, also successfully analyzed the implicit metaphors of practices in terms of how they align with psychological learning theories. Each theory might align with contemporary professional recommendations, yet Kirshner noted that learning progresses in fundamentally different ways in these commonly held views: Habituated learning develops incrementally, conceptual construction involves transformation of existing conceptual structures from peııurbations that arise out of reflective abstraction, and enculturation features discontinuity between prior patterns of paııicipation and new cultural patterns appropriated through cultural enmeshment. Kirshner hoped for a richer understanding of reform in mathematics education practice through the use of his "cross-disciplinary approach" that centered these distinctions. I am proposing something akin to Kirshner's approach. Arendt's categories highlight the activity in the classroom as constituting power relationships connected to broader sociological and political structures in society. How the mathematics is produced, and toward what purposes in terms of the community, is something that stands apart from more typical mathematics education research and reflection. Analysis using these categories of work, labor and action might dialogue with more commonly employed forms of research and practice.

Further Reading

Appelbaum, P. (2012). A game and a dare. In T. Kress & R. Lake (Eds.), *We saved the best for you: Letters of hope, wisdom and imagination to 21st century educators* (pp. 185–192). Teachers College Press.

Appelbaum, P. (2018). How to be a political social change mathematics education activist. In M. Jurdak & R. Vithal (Eds.), *Socio-political dimensions of mathematics education: Voices from margin to mainstream* [Conference session]. Monograph of the International Conference on Mathematics Education Topic Study Group on Social and Political Dimensions of Mathematics Education (pp. 53–73). Springer.

Pols, W., & Berding, J. (2018). "This is our world": Hannah Arendt on education. In P. Smeyers (Ed.), *International handbook of philosophy of education.* Springer International Handbooks of Education. Springer.

Hannah Arendt's Four Strategies for Unlearning

- Laughter
- Translation
- Forgiveness
- Dramatization

It is at least as important to "unlearn" as it is to "learn" in order to grow as a mathematical person (based heavily on Knott, 2015). We need to unlearn habits that get in the way of learning, to unlearn misconceptions and misleading assumptions, and also, subtly, to unlearn our forms of ignorance that are associated with things we know. Hannah Arendt used four strategies that we, too, can use to unlearn in our classrooms, to unlearn recognized trends and patterns—philosophical and cultural, to establish a praxis all one's own. To get unstuck.

Laughter

One of Hannah Arendt's most powerful strategies for developing the intellectual freedom to "unlearn" dominant philosophical and cultural prejudices is "laughter". Her model is the Thracian servant girl in Socrates' *Theaetetus*, who laughs at philosophers because they are so eager to know the things in the sky that they can't see what is there at their very feet. The Thracian girl represents a wordly perspective. Those who approach mathematics as professional "thinkers", fearful of the ridicule of the common man, need to unlearn through the laughter of innocence that can guide them into the worldly activities necessary for survival. This is not about hostility toward mathematics, but instead a joyful celebration of incongruences and absurdities, looking at mathematics without the seriousness of weighty respect. What this means is to revel in the paradoxes that pop up in mathematics, to laugh at the confusions that are so common, to enjoy the inconsistencies, rather than to insist on shoving mathematical facts and truths into the deadly prison of correctness. When can we play with more than one answer to a word problem by interpreting it in as many ways as possible? What are the extreme limits of values that we might find by changing the givens? Why do so many exercises seem absurd? When we get an answer to a mathematics question, what if we laugh at it by asking, what if this *is not* the answer? What else could it have been, and why? What if we change the original question into many similar ones by asking what-if-not about any particular aspect of it? We no longer revere the stodgy original questions as if they are weighty and important. We play with their changes. In doing so, we free ourselves to explore mathematics.

Translation

Translation was another important aspect of Arendt's work. She rarely if ever wrote in her native German, preferring to struggle with English, French, Spanish … because writing in a language that was not natural forced her to question what would be the best words to use. It also slowed down her thinking! Similarly, communicating in mathematical ideas, words, symbols and representations forces us to think about what we genuinely want to say. This changes the nature of writing in mathematics from "show your work" to "communicate what you truly want to say". It makes what we write very *important*. It is our presentation of our ideas to the world. What does our audience need from us in order to understand our feeble attempt to express our thinking? Classrooms that take this approach ask learners to create presentations of their work on a mathematics investigation, to explain their solutions, for different audiences. How do we write this as a little booklet for students in a lower grade? How about for our reading partners at the local senior center? What if we prepare this as a presentation for our families? Or, for students in a grade above us? They each require that we translate our thinking into very different languages, use very different kinds of pictures and information, and so on.

Forgiveness

Deeply significant in Arendt's writing, connected to the most horrendous traumas in history, is "forgiveness". It may seem almost wrong to try to apply it to our teaching. But then again, almost every student has been the victim of bad mathematics experiences by the time they come to us, so maybe we can start there. Things that have happened, Arendt notes, are irreversible. We can apply Arendt's writing about the Holocaust and other traumas to think about the traumas our students have gone through. We cannot undo the bad teaching and mistreatment of the past. We don't have to absolve those responsible for their actions. But we can "forgive them" in specific ways that keep the memory of their harm while not letting that harm prevent us from living in the present and future. Furthermore, misconceptions and misunderstandings that learners bring with them to our current lessons can take on this character of being forgiven, for the very reason that unlearning them through recognition of their harm is an actual act of growth and progress. Wrong answers are keys to understanding that such answers are the result of what would otherwise be excellent or correct thinking in a different situation, that is, they are glimpses of what we have in fact learned, even if we need to unlearn the habit of applying those procedures in these circumstances.

And, how about reflection on our teaching? How do we live with the unpredictability of our actions, our choices of what to say or not say, to do or not do? And when it seems to alienate or confuse a student, can we forgive ourselves? The best use of these past actions is to keep them in our memory, not to absolve ourselves of guilt, but to use them in ways that help us to better make decisions in the present and future. The beauty of forgiveness is that it interrupts otherwise automatic processes. For example, forgiveness enables actors to become freed from vengeance that encloses both doer and sufferer in the relentless automatism of the action process, which by itself need never come to an end. Within the space created by the interruption, forgiveness creates a new relationship that is radically different from what existed before. As something startlingly new, forgiveness is not conditioned by the wrong that provokes it, and it can therefore never be predicted. Arendt admits as much. She explains,

> forgiving, in other words, is the only reaction which does not merely re-act but acts anew and unexpectedly. Released from vengeful reactions, I can act in ways that are not predetermined or compelled by another's trespasses against me. In this sense, forgiveness is an unanticipated, uncaused and undetermined act; is it truly spontaneous. (Arendt, 1958, p. 241)

Arendtian forgiveness seems to take on a metaphysical stature; it appears to be able to change the nature of reality, undoing the irreversible. It acts against necessity, undoing what was done by releasing the doer from the deed.

Dramatization

Arendt came to see human existence as a stage. The job of a writer, she came to understand, didn't involve making an argument aimed to force the reader, through logic, to change his or her mind and come to accept what the writer had written. Her job, as she came to see it, was to spark a discussion in which readers were invited – indeed, expected – to take part. So, she would pepper her pages with quotations from other authors in order to create this discussion, this debate and this exchange of ideas in her text. The goal was to present a variety of ideas, perspectives and insights for the reader to sift through, evaluate, compare and contrast and, in his or her mind, synthesize into a new and personal understanding. The role of the teacher might be analogous, to create the mathematics in our classrooms as the stage upon which learners are invited – indeed, expected – to take part. The mathematics of the classroom challenges everyone in that room to change their minds about something. This is exciting! The mathematics in this classroom transforms everyone through thought and action from consumers of mathematics to producers of mathematics. Instead of just consuming the text of mathematics, they *think themselves into it*.

Further Reading

Appelbaum, P. (2009). Taking action – mathematics curricular organization for effective teaching and learning. *For the Learning of Mathematics*, *29*(2), 38–43.

Appelbaum, P., Stathopoulou, C., & Xenofontos, C. (2022). Mathematics education as dystopia: A future beyond. *Journal of Humanistic Mathematics*, *12*(2), 518–538.

Swanson, D. (2017). Mathematics education and the problem of forgetting: In search of research methodologies for global crisis. *Journal of Urban Mathematics Education*, *10*(1), 7–15.

CHAPTER 4

Math Joke Lists

– Top Dozen Math Jokes
– Common Homophones for Creating Math Jokes
– Polysemous Words with Special Mathematics Meanings

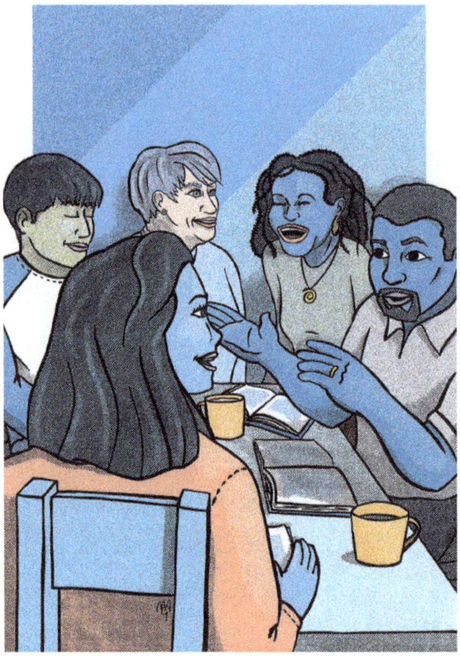

Top Dozen Silly Math Jokes

12. What's a math teacher's favorite season? SUMmer!
11. What do you call a number that can't sit still? A roamin' numeral!
10. Why did the student get upset when her teacher called her average? It was a 'mean' thing to say.
9. You should never start a conversation with Pi. It'll just go on and on forever.
8. Do you know why seven eight nine? Because you're supposed to eat three squared meals a day!

© PETER APPELBAUM, 2024 | DOI:10.1163/9789004688469_005

7. What do you get when you cross a dog and a calculator? A friend you can count on.
6. There's a fine line between a numerator and a denominator... But only a fraction would understand. Have you heard the one about the statistician? Probably.
5. A statistician got soaking wet trying to cross a river. He thought he could cross, because it was one-foot deep on average.
4. Why was the student confused when he went from English class to math class? Because he was taught that a double negative in English is bad, but in math, it's a positive.
3. Are monsters good at math? Not unless you Count Dracula.
2. There are three kinds of people in this world. Those who can count and those who can't.
1. ... and the number one silly math joke is ... Spoiler alert! - at the end of this section

Homophonics and Polysemicity

... are common sources of confusion and misunderstanding in mathematics classrooms. Learners often need to learn that a word has a special meaning when used in mathematics. In fact, this is such an important phenomenon that sociologists and linguists have a special term for it, the "mathematics register". Some people think of mathematics as similar to a foreign language, others to a unique dialect. Although many mathematical ideas are learned informally through everyday life, the special register is almost exclusively acquired in school, and is typically not spoken or written in most homes. Teachers need to exemplify and facilitate mathematics register fluency. There are many ways to accomplish this – whether you carefully model only the correct uses or encourage students to figure out meanings through context comprehension, or incorporate still other instructional strategies. Independent of your preferred approaches to the mathematics register, silly jokes highlight exactly the ways that the mathematics register differs from common everyday reading, writing, listening and speaking. Michael Halliday defines the mathematics register way back in the 1970s as:

> a set of meanings that is appropriate to a particular function of language, together with the words and structures which express these meanings. We can refer to a 'mathematics register', in the sense of the meanings that belong to the language of mathematics (the mathematical use of natural language, that is: not mathematics itself), and that a language must express if it is being used for mathematical purposes. (Halliday 1978, p. 195)

When we think about the mathematics register this way, access to language registers become associated with a student's social and cultural background – the opportunities they have had to live and experience life in situations that support the development of different language registers, shifting among them, and recognizing when they are in one social or cultural place rather than another. The German mathematics educators Hauke Strahler-Pohl and Uwe Gellert applied this idea to focus on the role of "gaps" in a mathematics classroom discourse between the mathematics register, the teacher's communication register(s), and the student(s)' various prior experience functioning in the various registers available and taking place in the flow of the activities. They showed how an observer can see the encounter as amplifying the lack of one or another fluency, and the implications for who has the opportunity to learn and who does not. Their work is grounded in another sociologist's theories, Basil Bernstein, who was primarily concerned with the relationships among language functions and forms of power and privilege in society. You can see how growing up with or without an awareness of homophonic and Polysemous aspects of language, and specifically mathematical examples, could interfere with learning mathematics. More importantly, though, is that a learner needs to develop a disposition to appreciate these aspects of knowledge and language to do well in school, which is tied to socio-economic status, immigration and language status, and other social and cultural dynamics.

The British mathematics education researcher Candia Morgan extends our appreciation of language and register by suggesting there is a great deal to learn about what is happening in a classroom by focusing on the contexts of communication taking place. First of all, she points us to the role of language happening. Telling a story? Explaining something or making an argument? Constructing a description? Giving a set of instructions for how to calculate something? What mode of delivery is taking place? If spoken, is this language moment establishing a new mathematical concept or procedure, testing students' recall or competence, explaining a task, developing a proof or a solution to a problem? Do the teacher and students both experience and recognize this language moment in overlapping and common ways? For example, do the teacher and most students comprehend that a story is being told? Or does the teacher think she is telling a story, and a lot of the students are hearing disciplinary instructions about correcting behavior, or are they trying to understand a new concept when there is none to be heard? What social cues are associated with each language action, and can a teacher and students develop a communal understanding of these, based on their outside of school cultural knowledge, and also grounded in some explicit discussion? A teacher can literally say, "Let me tell you a story. Sit back and listen to it first. Then I will tell it again and we will try to see if there is a mathematical relationship in the story that we can explore". With this sort of preparation, it is less likely that students would interpret the story as classroom management or a description of a mathematics idea to begin with. Candia Morgan also suggests that writing mathematically involves a great deal of experience with mathematical registers. Her work mostly advises researchers on how they might make use of students' written work, yet we can also think as teachers about whether or not the students are aware of their writing style as they write. What audience do they have in mind for their writing? Are they making lists of things to think about for themselves right now? Or, are they revising some messy written work to clearly communicate to another student in the class, who brings similar prior knowledge to what they read? Maybe they are currently writing for a highly skeptical person who is not in the class, or someone who they can assume does not really know anything about the ideas or any of the math terms they are routinely using. Writing for more than one audience is a version of register switching. Telling jokes is excellent preparation for this.

Jokes! Really silly jokes! Even jokes that make you groan!

Why would we tell them in mathematics classes? Isn't learning math important and serious? It turns out that "getting the joke" is a lot like "the secret behind a magic trick" or the "trick to solving a puzzle" ... and these moments of realization and understanding are tied to the subtle recognition that "you are in on it" and part of a special group of people who know why the trick works, know the secret, get the silliness that makes the joke teller want to tell the joke. Joke-telling taps into the processes of acculturation and enculturation (See Alan Bishop's Six Mathematical Activities in Chapter 1) that help learners see themselves as part of a classroom community, and who understand themselves as developing skills and authority – perhaps to make up their *own* jokes along the way. As young people grow up, they learn to laugh at silly jokes, groan and moan at "Dad jokes", and to gain attention by sharing jokes. They learn to enjoy the telling and listening to jokes. This is all part of acculturation in general. Math jokes are targeted mathematical acculturation, and simply ways to connect with already established cultural processes of acculturation. The awareness that some folks "get it" and others don't (yet?) is related to processes of *en*culturation, in which there is an authority or mastery associated with the hidden or secret knowledge (See Kirshner's Three Types of Activity in Chapter 1 for how puzzles can facilitate a sense of community belonging). Joke telling is an informal pedagogy aimed at both supporting a self-awareness of this authority and mastery, and a subtle, hidden curriculum linking mathematical knowledge with authority and mastery as a cultural value.

How do you invent a math joke? As the top ten list demonstrates, silly math jokes are typically constructed on one of two language characteristics. Homophones, words that sound the same but are unrelated, are plentiful in many languages. Because they sound the same, you can use one when the other makes what you are saying absurd. Polysemous words are also common. These are words that mean different things in different contexts. Instead of mixing up which word you are using, a joke that relies on polysemicity mixes up the contexts to create absurdity. Sample lists are included at the end of this section. You can probably think of many more to add to these lists.

By the way, the number 1 silly math joke is ...

1. What is a mathematician's favorite tree? – Geometry!

Further Reading

Halliday, M. A. K. (1978). *Language as social semiotic*. Edward Arnold Publishers.

Morgan, C. (2006). What does social semiotics have to offer mathematics education research? *Education Studies in Mathematics, 61*(1/2), 219–245.

Strahler-Pohl, H., & Gellert, U. (2013). Towards a Bernsteinian language of description for mathematics classroom discourse. *British Journal of Sociology of Education, 34*(3), 313–332.

Common Homophones for Creating Math Jokes

- Four and for
- Two and to
- Eight and ate
- Nine and nein (the German word for "no")
- Acute and cute
- Pi and pie
- Sign and sine
- Add and ad

Polysemous Words with Special Mathematics Meanings

- Angle: formed by 2 rays joined at a vertex, versus Angle: a viewpoint, perspective, or actual tilt
- Constant: a number that doesn't change (not a variable), versus Constant: occurring continuously over a period of time, or a situation that doesn't change
- Digit: any of the numerals from 0 through 9 used in a base-ten representation of a number, versus Digit: a finger or toe
- Even: can be written as 2n, where n is an integer, versus Even (if): make flat or smooth, or to show extreme degree
- Function: a special kind of mapping in mathematics, versus Function: a feature, purpose, or way of creating an outcome
- Gross: 12 of something, a dozen, versus "ick!"
- Key or Legend: List of labels on a graph or items in a map, versus Key: what you use to open the door, or Legend: a great story
- Matrix: an array of numbers, symbols or other values, arranged in rows and columns, versus Matrix: a system or environment in which somethings exists or develops, a mold for casting a print type
- Or: a non-exclusive potential alternative, versus Or: an exclusive, different attribute or category
- Orientation: the angle of an object relative to the axes in a graph or reference frame, versus Orientation: training for new members of a group, or feelings and beliefs
- Power: the number of times a base is multiplied by itself according to an exponent, versus the capacity to influence others
- Real: a special type of number – which is not more real or not than any other type of number, versus "existing in reality", actual. (similar distinctions should be made for complex, rational, natural, etc., numbers)
- Root: a point where or value for which a function is zero, versus Root: a type of vegetable or a source of something
- Series: the sum of the terms of an infinite sequence, versus Series: a set of related television or radio programs, especially of a specified kind
- Similar: having the same shape but not the same size, versus Similar: resembling without being exactly the same.
- Table: mathematical information organized in columns and rows, versus Table: a piece of furniture with a flat top and legs
- Yard: a unit of measure equal to 3 feet, versus Yard: the area around a building

The original idea for these lists was that they are resources for inventing silly math jokes. Posting these lists on a classroom wall and adding to them as a class over the course of a school year can lead to regular sharing of new jokes. Word wall cards, popular in elementary language arts, are enhanced as math joke word cards. You might set aside 5 minutes weekly or monthly as "math joke time" or more casually share and appreciate jokes as they spontaneously arise.

It's important to explicitly teach specialized mathematics meanings. Research demonstrates that students will otherwise ignore the math words and only pay attention to the mathematical objects and just guess at how to combine or manipulate them. If asked, "Primo has 32 soft pretzels to share with 8 team-mates; how many pretzels will each get?" there are a host of unspoken possibilities about sharing equally, let alone thinking about the meaning of the words. Declan might just think, "Well, 32 is bigger than 8, so I'll subtract and say 24". Shanda might read, "$f(1/(1+x)) = 2x + 12$. What is $f(1)$?" and think, "I don't know, maybe I can put in $x = 1$ and get that $f(1/2) = 1 + 12$, but that doesn't help me since f could be any function so this value is not all that useful", and just give up. 5th-grade teacher Kathleen Palmieri took on vocabulary as reflective teaching project and now advises that learning mathematics vocabulary is much more than studying definitions. She describes designing classroom activity over an extended period of months to promote an "architecture of learning" in which concepts are explored and a pathway of understanding is established before a mathematical term can be attached to establish true meaning. Palmieri takes a "topic walk" with her students before beginning a new unit, to identify special terms and to create a vocabulary list. She intentionally collaborates with the students to decide what words will be in a classroom glossary, as well as finding, discussing, and agreeing on definitions and examples to include. Students create and compare graphic organizers for special terms, in which they surround a word or phrase with examples of its uses in sentences, related words, and important non-examples. Students create special word posters, and tally how many times a word of the week is used throughout all activities. Key in her approach is how students are included as active designers of their vocabulary building. Students begin with their own language to describe and contextualize concepts and relationships. She deemphasizes definitions as correct or wrong, by generating conversations around the various ways that terms are used to make meaning. Each interaction with language supports individual understanding of the situations being investigated, rather than functioning as a learning objective.

Swedish researcher Ulrika Ryan collaborated with Greek co-researcher Anna Chronaki to observe a classroom in southern Sweden in which a young student made a somewhat joking side comment about other students' body parts. The moment highlighted the many ways in which spontaneous jokes are openings to important comprehension during the complex and multifarious nature of mathematics classrooms. First of all, they note, jokes are ways that humans cope with uncertainties, incongruences and complexities. Hearing a joke might signal that a problem is at a good level of challenge, or that students are on the verge of unproductive frustration. Certain jokes are at the expense of others, and in the case these researchers studied, the specific joke had this possible interpretation, establishing a feeling or superiority or control over the young men in the class by the young women. However, a more interesting aspect of this joke was how it demonstrated a fairly sophisticated level of understanding of angle classifications, and at the same time a commentary on why precision in measuring might be important. Observing the subsequent reception of the joke by the small working group, the researchers demonstrated how a teacher can see the "joke-event" as establishing comfort with each other, rather than gender competition, fluency with mathematical reasoning, as well as how they work together or in parallel as members of a group, paying attention to a joke, or not, as they demonstrate focus on the assigned task. It also revealed the students' general informal relationship with precision in school mathematics, as they joked about measurement both as establishing a precise result and as being difficult to carry out with precision. It is just such a rich comprehension of mathematics as both straight forward and complex at the same time that the homophonic and polysemous characteristics of mathematics terms contribute to apprenticeship in mathematical reasoning. As Ryan and Chronaki also point out, spontaneous jokes reveal a space of double tension that is at the heart of a great deal of mathematics. On the one hand, learning mathematics and thinking mathematically about things requires conforming to norms and expectations, conventions about how things are done and interpreted. On the other hand, breakthroughs in understanding, pathways to solutions, or the invention of new mathematical concepts often take place by challenging norms and assumptions, or by changing the question (See Brown & Walter Problem Posing in Chapter 8).

Further Reading

Ingran, J. (2020). *The words we use*. Association of Teachers of Mathematics. YouTube Video. https://www.youtube.com/watch?v=goWYJMLTYfk

Palmieri, K. (2022). *How we pumped up our math vocabulary study*. https://www.middleweb.com/47850/how-we-pumped-up-our-math-vocabulary-study/

Ryan, U., & Chronaki, A. (2020). A joke on precision? Revisiting "precision" in the school mathematics discourse. *Educational Studies in Mathematics, 104*(3), 369–384.

CHAPTER 5

Math as Art Lists

- Josiah McElheny's Three Ways to Use Models
- Susan Sontag's Alternatives to Representation
- Sianne Ngai and the Aesthetics of Students' Contributions
- Art Projects for Math Classes

Josiah McElheny's Three Ways to Use Models

- Architect
- Scientist
- Artist

How we use representations to pursue our activity, and to communicate and share with others, can be characterized in different ways, leading to different goals or purposes for the activity in the room. Glass-blowing artist

© PETER APPELBAUM, 2024 | DOI:10.1163/9789004688469_006

Josiah McElheny describes three ways that artists can use models in their work, based on heuristic stereotypes. If we think of mathematical activity as artistic activity, and what learners produce with mathematics as art, then the extension makes perfect sense.

Architect

An architect uses their (scale) model to convince a client, their audience, to believe in their project.

Scientist

Scientists represent reality, analyze relationships among elements, and test hypotheses with models.

Artist

McElheny suggests that artists challenge assumptions and raise questions with their work. Their work can function as a model that provokes, wonders, or offers alternative ways of seeing and thinking.

Models and Representations

Models and representations are central to mathematical activity. Physical materials, diagrams, charts, symbolic representations, exemplary problems of a type, etc., dominate mathematics teaching and learning. Much of our pedagogy rushes to representations. We want pupils to become very good at using them to model concepts and relationships. This desire, indeed, seems to be at the heart of mathematics as a living, growing discipline: pupils who can move from one model to another are usually taken as understanding the mathematics. It seems that the very act of working with representations as models might actually characterize mathematics. We often assume we know what we mean by representations and models, yet there are multiple ways to conceive and apply them as researchers, as developers of curriculum materials, as teachers of mathematics, and as mathematicians. In particular, there is a widespread assumption that models are used to create a picture of a real-life situation to facilitate problem-solving, or that a representation within one mathematical area might be employed as a model of relationships in a previously disconnected branch of mathematics. To presume that models are tools of problem-solving, for example, contains the further expectation that we would explain how to use models to someone who does not yet know how to do this. What if the people with whom we interact are using models for other purposes? If we assume their ignorance, then we are reifying our own ignorance of the ways they are modelling and using models.

The point is not to caricature professions, but to use dominant stereotypes to generate types of practices. Each stereotyped practice applies metaphors that, in turn, represent assumptions, values, fears, desires, structures of discourse, and so on. We might imagine teachers or pupils of mathematics similarly employing models and the representations within them with different aims at different times and, in this sense, acting in ways analogous to different iconic conceptions of representation in general. For example, a teacher or pupil might work as an architect-mathematician, scientist-mathematician, or artist-mathematician, at different times during a school day, in order to achieve various goals, just as they probably do in their everyday life when the situation calls for these orientations. This is not to say that such supposedly non-mathematical approaches would somehow rescue mathematics, nor that students might find working like an architect, scientist, or artist more relevant or motivating than working as a mathematician. We are using these professions metaphorically to describe the experience of social interaction and implicit epistemological conceptions of representation within a cultural discourse.

One might employ blocks, pictures, or equations as part of an argument to convince others of a conclusion or the reasonableness of a result, metaphorically acting as an architect convincing clients with a blueprint or scale model of a building. Or one might invent a way of representing a system or set of relationships, and then proceed to analyze the representation for what it further implies regarding these relationships, much as a scientist uses models to better comprehend the relationships among natural phenomena. Architect-mathematicians might employ a construction by straightedge and compass to convince another pupil that the perpendicular bisectors of the sides of a triangle will intersect in one unique point; we might create an animation using dynamic geometry software or create a series of logical statements that include details about these parts of a triangle in algebraic symbols.

If learners graph parabolic data and use the graph to study parabolic functions (noting various characteristics of the types of relationships that are demonstrated among the variables that are depicted by their graph), they could be metaphorically labelled as scientist-mathematicians; similarly, drawing a picture of a person and their shadow from a lamp-post as they walk, marking items that could be measured at discrete distances from the lamp-post along the ground, a scientist mathematician would then use the picture to generate a chart or graph from which they could further clarify particular aspects of the relationships involved, i.e., proportionality between heights of the light source and the person walking, length of shadow, distance from lamp-post, and so on.

Artist-Mathematicians

In McElheny's (2007) sense, artist mathematicians seem rarer in school. Yet, I believe that this way of using models is more central to the work of mathematicians than architect or scientist-mathematicians, at least, more at the heart of what the mathematical experience is all about. Conceptual artists do not use models to convince others or to solve problems. McElheny's artists use models to create new worlds, imaginary spaces of learning outside of time and space. The models become "proposals" – invitations to come and play and explore ideas. The other kinds of models drag us down into realms of accuracy, correctness, and explanation: "Is it a 'good model'?" is too often taken to mean, "Is this model a true replica of the real world?" Such models are tossed aside as soon as they fail to live up to the demand that they precisely mirror nature. McElheny refers to Isamu Noguchi's playground designs. Noguchi's model for a UN Playground in New York was never built, but the models became well-known works of art in museum exhibits. Used in this way, the models provoke us to imagine what playgrounds could be, to think about our assumptions about what a playground should or should not be, and to question the decisions that we generally make about what might take place in a playground. The artist's model differs in value from other models, it seems; perhaps it raises the potential for all models, regardless of their surface intent, to carry this provocation. Not tied to true mirroring, the pleasures and utility of an artist's model are in itself, and in the way it questions the very act of modelling. Conceptual art leads a person interacting with it to reflect on the process of making the art and the concepts that are invoked with the art. It is in this sense that I believe McElheny's artist is like our student mathematicians, contributing to a living, growing discipline. Noguchi's playground is, simultaneously, a potential playground in the world and a playground of the mind, generating all sorts of experiments and questions in time and space distinct from our direct experience of our world, yet maybe influencing that world through the ways we might act in the future, even more than a playground might affect the lives of whoever had played in it. It is in this sense that the model becomes a "sculpture of utopia" – an ideal conception that does not exist in the world. A mathematician, artist or practitioner of any kind who establishes this sort of standpoint on his or her use of models evokes what Brian Rotman once called a meta-subject – someone compelled to consider the relationship to the act of idea generation or artistic invention itself; the work provokes reflection on the meaning of the work with the very action of constructing the model. For mathematics education, model creation is critical, because it enables the teacher and student to talk about the specific purpose of mathematical ideas, as well as to reflect on the processes of idea-development: it instantiates those aspects of learning and concept creation most difficult to address in ordinary classroom practices. Both teacher and pupil examine and discuss what is typically left to chance: processes of creating, representing, and modifying ideas.

Further Reading

McElheny, J. (2007). *Artists and models*. Presentation at the Museum of Modern Art, New York.
Rotman, B. (1993). *Signifying nothing: The semiotics of zero*. Stanford University Press.

Susan Sontag's Alternatives to Representation

- Parody
- Abstraction
- Decoration
- Non-art

Representations can be thought of as the actual material and content of mathematics. I mean here simple things like numerals to represent numbers of things; drawings of shapes to represent ideal geometric relationships; fractions to represent parts of wholes, proportions, and ratios; equations to represent functional relationships; letters to represent variables that may take on different values; base-ten blocks for arithmetic operations; drawings of rectangles or circles for fractions and ratios; graphs that visually represent algebraic equations, modelling mathematical concepts and relationships.

One interesting point to then consider is that art historians and critics sometimes suggest representation is not always the aim of art; in fact, for some, representation *violates* art. What might learners as conceptual mathematical artists do, then, if they are not primarily practicing forms of mathematical representation? What would it mean for youth who are learning "stuff that many adults already know" to be *artists* – creators and producers – when we seem to want them to consume and use mathematics instead? The critical issue turns out to be the way we make sense of the "art". Philosopher and cultural critica Susan Sontag (1966) wrote of the "highly dubious theory" (p. 10) that art contains content. When we take art as containing content, we are led to assume that art represents and interprets the world, and that these acts of representation and interpretation are the essence of art itself. Similarly, we often imagine the mathematics curriculum as content and move quickly to the assumption that this curriculum represents and interprets. This list is very radical and weird for most of us: It is asking us to imagine that the mathematics might be something other than "representing something or someone". That usual way of thinking about representations in mathematics classrooms makes the mathematics-art and curriculum into articles of use, for arrangement into a mental scheme of categories. What else could art or curriculum *do*? Sontag actually has suggestions. She suggests several things: "To avoid interpretation, art may become parody. Or it may become abstract. Or it may become ('merely') decorative. Or it may become non-art" (p. 10). Parody, abstraction, decoration, and/or non-art are four tactics for art and for the mathematics curriculum.

Mathematical work is not always done within the framework of McElheny's architect or scientist, representing things in the world, creating a picture of something outside of itself. More importantly, there is a way in which even the work of an architect or scientist relies on the kind of practices that McElheny calls an "artist", or eventually leads to this artist-like way of working. The artist-aspects of such work are prior to, anterior to, or independent of the work of the architect or scientist. Given the extensive, wide-ranging research on modelling and models in mathematics education, I suspect my argument is challenging. Most of us already have a working understanding of models and representations, and most of us employ a number of assumptions about them in our own work. I am asking us to reconceptualize some aspects of mathematics that are very fundamental to what we do and, in the process, opening myself to the criticism that I am doing what I ask us not to do: explaining what needs to be known. It may take a great deal of effort to explore this together. I have found recent conversations collapsing back into a discourse that presumed, within a pedagogic stance, a scientist-mathematician framework that could not allow the comparisons that this list offers. If a teacher presupposes that most activity in a classroom should have the purpose of solving a problem or of practicing methods of solving types of problems, then that teacher is going to think of models only in the way of a scientist-mathematician, since the main purpose of models in such a classroom is to accurately present a mathematized analogy for the situation occurring in the problem to be solved. This means that such a teacher will have to reorient themselves in order to take advantage of what I am saying. They might, for example, begin to introduce new types of activities in their classroom, during which pupils are not solving problems or practicing methods of solving problems; such activities would involve pupils comparing and contrasting models, with no intention of using the models to solve a problem; or taking a model for one mathematical concept and absurdly using it to demonstrate a seemingly unrelated mathematical set of relationships.

A teacher who values problem-solving for its clear selling points on the usefulness of mathematics will also find these ideas challenging. In some ways, the artist-mathematician seems to be less concerned with the real world or with the application of their ideas. This distancing from reality is not really the case, but the issue deserves some attention. Zbiek and Connor (2006) offer modelling as an activity that replaces motivation, providing a context for learning the mathematics itself; this, too, is different from the artist mathematician, who might, in fact, care more about the potential of a model or representation than the mathematics for which it was initially brought into use. The history of mathematics is filled with those, such as Hardy (1940), who have extolled the virtues of pure mathematics, whose efforts seem to exist independent of the practicalities and necessities of an architect, or whose work has no intended scientific application. Even a pure mathematician, however, employs the habits and skills of an architect when convincing others of their conclusions or proofs, and those of a scientist in elaborating and evolving models of systems of relationships.

At one point, I tried to write this book with numerous examples of classroom practices. However, as I finally understood, it is not the lesson plans or curriculum materials that are the critical element. What matters is precisely what is not in those plans or materials: the orientation to mathematics as sculpture of utopia, in which the relationship with the models and representations is analogous to conceptual artists working with found objects. Rather than refining how well a model represents reality, teachers could employ a series of exercises adapted from theater activist Augusto Boal's (1995) *Rainbow of Desire*. Pupils would critique models as if they are works of mathematical art. Many lessons would look ordinary on the surface, yet the students would routinely juxtapose, appraise, analyze and create models; place models in historical context; and use models to provoke emotions, such as joy, nostalgia, outrage, or constancy. Each representation – every picture, diagram, chart, graph, equation, etc. – would, in such classrooms, always be taken as a potential model of many concepts and relationships. And this notion that a particular representation is always potentially many different models all at once is important, because such different uses of the representation as a model can be discussed, explored, applied, critiqued, modified, etc. Seemingly unrelated concepts and relationships modelled by a common representation are drawn together into new worlds of similarity and difference within classroom conversations and pupil investigations. This all occurs outside of the time and space of the literal modelling processes that make up the usual or traditional focus – in other words, within a utopian place, both inside and outside of time and space.

In one educational environment outside of the usual time and space of school mathematics, an after-school "math circle" with secondary school students in a high-poverty, high-crime, low school-performance neighborhood of

Philadelphia, youth used statistics to design a basketball "dream team". A traditional approach would have led the students through an explanation of existing sports statistics and the practice of using them within a well-known model of basketball skills to deduce what many sports analysts already know about the perfect team of players. Instead, university pre-service teachers facilitated the investigation, which was designed by the high school students, helping them pursue a series of experiments in a gym at the local church to determine particular skills and statistics upon which their dream-team would be based. Multiple models and the dependent importance of one statistic or another and how they might apply to the dream-team design were proposed, tested, and compared. In their discussion of which models to employ, students needed to consider how to represent their model for themselves and for their colleagues. In a primary school math circle, repeated addition on a calculator led students through a series of efforts to model the phenomena that arose, using a number line, a 100s-chart, a collection of tessellation tiles, and a circular representation of modular arithmetic. In many classrooms, students would have been shown how to use the models as part of being guided to deduce the patterns in the units digits of such repeated addition, and the lengths of those patterns, in order to learn particular ways to seed repeated addition with a starting number and a constant adding number for specific outcomes. The university students facilitating this math circle encouraged a small group of children who happened upon repeated addition on their calculator to share what they had noticed with the rest of the group. Discussions emphasized differences among the types of models for what they revealed and obscured. Similarly, attention to sketches of an African village's architecture, shadows cast by a sculptor's collections of human trash, or a graph of recent income distribution would help students understand the kinds of relationships and questions that can be examined and posed as much as such attention would help them to practice specific skills.

Further Reading

Appelbaum, P. (2012). Mathematical practice as sculpture of Utopia: Models, ignorance, and the emancipated spectator, *For the Learning of Mathematics, 32*(2), 14–19.

Boal, A. (1995). *The rainbow of desire: The Boal method of theater and therapy*. Routledge.

Hardy, G. H. (1940). *A mathematician's apology*. Cambridge University Press.

Zbiek, R. M., & Conner, A. (2006). Beyond motivation: Exploring mathematical modeling as a context for deepening students' understandings of curricular mathematics. *Educational Studies in Mathematics, 63*(1), 89–112.

Sianne Ngai and the Aesthetics of Students' Contributions

- The Ugly
- The Cute
- The Zany
- The Interesting
- The Gimmick

Once we believe in our students as artists producing "art" we can explore our reactions to what they produce – the things they say, the things they make, the things they use to make their mathematical "art". What feelings accompany the moment we see or hear a student's work? Do some students have what we might call a "style"? Aesthetics is that branch of philosophy concerned with the nature and appreciation of beauty, especially in art. Some teachers study what philosophers have said across the centuries about beauty and appreciation, in order to better understand their students' conversations, explanations, written responses to problems, and so on. However, we can also recognize with the scholar Sianne Ngai (2005, 2012, 2020) that aesthetic judgements always begin with feelings, that is, fine gradations of affect inspired by an encounter – with an object, event, whatever, held up and treated as something to consider – feelings only later possible to articulate and justify with an explanation. How often is a student's idea experienced as "cute" or something like cute? A sentimental attempt to please us? How frequently are we disappointed in student work? Envious, irritated by what they are doing, anxious about what they are about to do? Ngai suggests these "ugly" feelings are associated with situations in which action is blocked or suspended, and that they are increasingly common symptoms of our cultural moment: we feel them a lot, not only in our teaching, but everywhere.

Ngai offers the terms in this list as especially useful for understanding how our aesthetic experiences (of our students' work, including how they participate and produce mathematical thinking, conversations, distractions, and so on) have been transformed in the "hypercommodified, information-saturated, performance-driven conditions" of life in the 2000s – that is, the conditions of life which define teaching and learning in our historical moment of standards and testing and outcomes-based practices. She is interested in these terms as categories because she believes that the ways we respond to art can help us understand important non-art aspects of our culture. I recommend trying this! We can think about the ugly, the cute, the zany and interesting, and the gimmicks, to help us see how we and our students are using and processing information, how we work mathematically, how we play with mathematics, how we trade, buy and sell our efforts and our ideas for things we value. Ngai suggests these categories of feelings are so powerful because they are so easily unnoticed.

Book critic Rebecca Ariel Porte (2012) points out that Ngai's categories make it possible to see beyond good/bad evaluations of work or forms of participation or contribution, and to do the more serious exploration of how we arrive at our value judgements in the first place. We can use this idea to do more powerful assessment, both formative and summative (See the Alice Pitt, Susan Ohanian, and Jo Boaler assessment lists in Chapter 3). Those "cute" things that learners create or do in class share a lot with other cute things in our lives. They are adorable like frog-shaped sponges, panda cubs, or some *anime* characters. (Ngai points out that cute is a mid-nineteenth century shortening of the word acute, shorn of its sharp angles.) To judge something cute, according to Ngai, you first need to feel your own dominance in relation to it. And surely teachers have an authority that puts them in a dominant position to judge student work or forms of participation. Ngai's sense of dominance might be a kind of desire to protect the cute object of your gaze, like a parent might, or in a darker version, to destroy the cuteness to promote maturity. The main relationship, however, is one in which the cute stuff that learners do is powerless to the teacher's actions and how the teacher "uses" it. Many times, we use student work or things they say without considering a student's perspective on that use, and this reduces it to cuteness.

Interesting

... is an aesthetic judgement so mild and commonplace for Ngai that it hardly seems to qualify as a judgement at all, but that's her point. If it is an aesthetic judgement, then it is one that emphasizes the ambiguity of what we are experiencing. Ngai explains that "interesting" begins with curiosity and wonder, which is why I want to seek more of this in my own teaching: inquisitiveness is something I value and want my students to value as well. However, the interesting can end up meaning something like "merely interesting", just as much as it might be a statement describing "good mathematics". For me, a conversation about whether a mathematical idea is "interesting!" or "merely interesting ..." would be worth our time in our class. Interesting things do always run the risk of being experienced as boring. Yet, the "good" sorts of interesting are like those uses of art to provoke questions that Josiah McElheny describes, opening our imagination. Similarly, when we are interested in what some students are doing and why, this means that our experience of teaching these students is one that piques our imagination.

Zany

… conjures images of the unconventional, the surprising and unexpected. Strangely humorous forms of mathematics and zany questions from students challenge the boundaries of work and play. We recognize them as a sort of virtuosity but we're afraid of them at the same time because they move so quickly and erratically that we can't predict them. They suddenly shift the focus of a lesson at any given moment. We feel like the straight man for a zany comedian. Zany moments and zany ideas are wonderful at a distance but feel dangerous up close when they challenge our expertise, distract us from our plans, or ask us to be zany ourselves. If we are zany, we can get lost in that zaniness, exploring the wacky and unconventional. Sometimes this is what problem posing feels like, as we enter unpredictable investigations when we ask, "What if not?" (See Brown and Walter's problem posing list in Chapter 8) Yet, without the zany, we wouldn't invent new things, ask exciting questions, or provoke totally new insights, and this is perhaps the best of the problem posing contribution to mathematical experience. Sometimes, the math jokes are so silly that they conjure this type of aesthetic encounter. (See the math joke lists in Chapter 5) The zany is fun, and entertaining, and balances the merely interesting, pulling it out of its boring elements, turning the cute into the truly interesting.

The Gimmick

In a sense, each of the lists in this book is an example of Ngai's most recent aesthetic category, the "gimmick". A gimmick can get to think differently about teaching and learning mathematics. Ngai believes that "the gimmick" is fundamental to our contemporary culture, and in this respect, lists might just be the trick to making our teaching "interesting" – in a good way. Ngai defines a gimmick as something aesthetically suspicious because of the way in which it seems to be working either too hard or too little. Hopefully, the lists in this book are really helpful, so that they are genuine tools and not merely gimmicks. Yet our experience of a gimmick's impoverished form triggers evaluations of excessive or deficient labor that highlight its false claims to value. A gimmick is flagrantly unworthy yet perversely attractive: a worksheet; a standardized exam; a lesson plan on the internet, complete with reproducible materials. They appeal to us, yet they hardly teach or offer meaningful assessment. They work hard and do very little teaching. We want what they promise, yet their form makes them disappointing. Ngai is concerned with gimmicks because she uses them to explore how our capitalist system's weaknesses are revealed through gimmicks – they promise value and do not deliver on their promises. Like a laugh track on a television sitcom, a gimmick is both under- and overperforming at the same time. Supposedly time-saving, performance-enhancing devices are gimmicks: in school mathematics, standard algorithms that replace conceptual understanding, graphing calculators used for simplistic tasks, "think-pair-share", "convince yourself, convince a friend, convince a skeptic", "growth mindsets", so-called learning styles, and so on.

The idea of the gimmick doesn't tell us anything we didn't already know. But it makes clear how much the everyday aesthetic language of the ugly, the cute, the interesting and the zany can help us make sense of how we are feeling. This aesthetic language might also be a gimmick we can share with our students! Try asking them to use these terms to judge the mathematical ideas and strategies they are learning about. When is a zany math concept or a cute algorithm interesting? Why is it so easy to slip into boredom? If Ngai is correct, then these categories of experience capture our current cultural moment and connect us to our feelings. They would help students understand themselves and the world, through their judgements of mathematics and mathematical thinking.

Further Reading

Ngai, S. (2005). *Ugly feelings*. Harvard University Press.

Ngai, S. (2012). *Our aesthetic categories: Zany, cute, interesting*. Harvard University Press.

Ngai, S. (2020). *Theory of the gimmick: Aesthetic judgment and capitalist form*. Harvard University Press.

Porte, R. A. (2012, October 14). *The zany, the cute, and the interesting: On Ngai's "Our Aesthetic Categories"*. Los Angeles Review of Books. https://lareviewofbooks.org/article/the-zany-the-cute-and-the-interesting-on-ngais-our-aesthetic-categories/

Art Projects for Math Classes

- Classroom Studio
- Exhibitions
- Public Math Provocations
- Visual Communication Challenges
- Soundtracks
- Performance Art Development
- Mathematical Docents & Critics

Building on the Josiah McElheny and Susan Sontag lists, mathematics as the art "that builds communities characterized by joy, a can-do attitude, and the courage to act on their convictions" might take on any of the approaches listed here. First of all, if we start with a belief in each class participant as a "mathematical artist" creating with mathematics, creating their own version of mathematics, sharing in the aesthetic traditions and forms of mathematics, innovating within those traditions, breaking new ground in their invention of ways to be a (mathematical) artist, then one way to support these ways of being is to pattern our learning spaces on the ways that artists have trained and developed their craft, their senses of judgement, and the touchstones to which they connect as they become their own uniquely creative selves.

Studio

Art schools often use a studio structure, in which students work on individual or collaborative projects in parallel. Periodic pauses allow one student or team at a time to present work in progress for feedback, critique, and emotional reactions from the others, which they can bring back into the developing project. The studio teacher poses occasional exercises designed to build techniques and skills that they identify as potentially useful to the students in the class, based on their observations of their process and subject choices. This is a modern version of the historic role that apprenticeship in an *Atelier* (workshop) enabled. The *Atelier* was typically oriented toward apprentices learning to copy exactly how the master worked, until they were "ready" to produce a "masterpiece" worthy of their rise to the status of running their own *Atelier*. A more open-ended studio or a workshop designed to train in specific skills is a matter of authority and knowledge. There is of course room for both approaches, a teacher might vary between each at different times, yet each encourages a sensibility guided by the view of the student as evolving mathematical artist.

Exhibitions

... grow out of the work that artists are creating. Any student work might become sharable in an exhibition. They are sometimes "private", shared only with each other, or special invited visitors. Other exhibitions can be more public, in public venues such as school lobbies, hallways, community centers, sponsoring local businesses, and so on. A lesson might simply ask a few or each student to display work in progress, or more polished work, to which the class visits and reacts. Feedback sheets near each student's work can collect reactions. A more formal rotation among students in a studio-type critique session serves a similar purpose. Students need guidance at first in sharing honest emotional reactions, supportive notes about mathematical technique and craft, appreciations, and suggestions for improvement. Other exhibitions are more public, set up like a museum for outside-of-class visitors. One student, a small group, or the entire class will at times serve as the curator(s) of the exhibition, deciding everything from overarching themes to design of the presentation space to the flow of visitor traffic and explanatory materials available for facilitating interaction with the mathematics on display. As each type of exhibition is experienced, subsequent reflections on the successes and ideas for the next can make each new exhibition more exciting and potentially more educative. Visits to professional and math-class art and mathematics exhibits, museums, and other public event spaces can provide inspiration for the class.

Public Provocations

... might look like an exhibition in a surprising venue – for example, an exhibit of work in several stages of progress at a shopping mall, public library, town park or bus stop could be organized around the processes of problem solving within a specific mathematics topic. They might also look very different. Students might create posters for public transportation that use mathematics to shock commuters into understanding a social issue. Students might set up a table in the school cafeteria or in nearby nail salons to offer assistance in mathematical problem solving (filing tax returns, creating a weekly budget, understanding news articles about current political issues that use mathematics ...). A mathematics "party" or "picnic" can share completed work, or encourage visitors to collaborate on their own. Public collaborations might lead to a permanent mural, a new playground design, a parade, or even something the organizers haven't yet planned.

Visual Communication Challenges

... can be the content for an exhibition or provocation, or components of more traditional classroom lessons. Explaining an idea or mathematical relationship on a poster without using words or numeral symbols, convincing others of an argument in only pictures, and communicating your mathematics idea silently, in gestures only, are all examples of this approach. Theater tableaux are a more established form for this. To create tableaux, a small group of students thinks of how they can become a series of three to four frozen collections of statues, so that the sequence of each tableau one after the other creates a mathematical story. Mobile phones can create a series of photos or video that preserve visual communication challenges to be viewed by a larger audience beyond the moment.

Soundtracks

In contrast, soundtracks move away from visual representations toward a more metaphorical experience of mathematics. Students can compose their own soundtrack out of effects that they make themselves or collect digitally, or share excerpts from music recordings, communicating how they feel about a mathematics idea or a challenging mathematics investigation. They can use a mathematical pattern or other mathematical idea as the inspiration or structure for a sound composition or for organizing how classmates create sound combinations to perform together.

Performance Art Development

Yet another direction for mathematical artists is to use mathematics to practice and perform an art form for which there is a symbolic notation system. Dance choreography uses symbolic notation, and working with this notation shares many characteristics with mathematical systems. Music notation is a go-to for many teachers. However, a more fruitful approach to music is to have students invent their *own* notation systems that represent what *they* want to remember and reproduce in future performances of a musical idea they create, music compositions, etc. How do they want to represent pitches, dynamics, tempos, timbres, sound sources, timing of sound making, changes in any of these quickly or slowly? Once they have a collection of symbols, the power of "algebra" is available to them: what combinations of these characteristics do they feel lead to other combinations according to their creative spirit? This is a function of sound variables! Collecting functions like this establishes a world of music within which they can work. Juggling, braiding and bell ringing all share a mathematical system that is easily learned. Once students have practiced some basic patterns and learned how the notation system works, they can either work within the system to invent new patterns to practice and perform or introduce additional effects and combinations of their own into the conventional system, and work with them to develop new performance ideas. Origami artists use visual diagrams to communicate types of folds. The geometry of foldable and impossible origami designs can lead to students working with the visual crease patterns to invent their own designs. Even more exciting are the ways that standard "bases" in origami mirror what college mathematics majors call "linear algebra", in which an initial base set of folds creates a space of functions and algorithms for specific foldable designs. In general, any passion can be explored in this way, establishing a mathematical set of representations for transformations and actions on objects of interest, real or imagined. Once approached in this way, the mathematical interaction between processes and representations generates an "algebra" of those objects of interest, upon which artists can invent explorations, can be explored first "in real life" and then represented in the notation system, then investigated in the notational representations in order to discover possibilities that can be tried out "in real life". Some rich possibilities include recipes, cooking and baking; fashion, scarf-tying designs, and outfit coordination; sports playbook options; and color mixing for painting or drawing.

Mathematical Docents & Critics

The last item in our math as art list supports students as docents and critics. Docents typically lead tours of an exhibit, for example as volunteers at an art museum or historic site, sharing entertaining stories and facts that visitors can use to more fully engage. Mathematical docents naturally take a special mathematical approach to their tours. To prepare their tours, they spend time at the venue, open to any and all mathematical ways to experience the space and its content. They allow the art on display, facts that can be learned about the place or the material, and so on, to provoke mathematical ways to engage (For ways to explore art, things in specific places, and historical aspects of a location, see Alan Bishops Six Mathematical Activities and Tamsin Meaney's List of Bishop's Activities in Chapter 1). A docent spends a substantial amount of time in open-ended consideration of possible mathematical questions they might use to interact in the venue, and eventually puts together a tour plan, using the mathematical stories, facts, and challenging questions they could share in their tour (See the Problem Solving and Posing Playbooks in Chapter 8, and especially Brown and Walter's *Art of Problem Posing* for strategies of initial docent tour creation).

Mathematical critics use their mathematical methods of inquiry and mathematics knowledge to describe what they see. Similar to docents, they start by visiting a place, viewing a film TV show, dance concert, or engaging with any work of art, video game, garden, city transportation system, etc., collecting notes on how numbers and functions, shapes and transformations, data and statistics, information systems and logic lead them to understand, feel, and wonder about whatever calls their attention. Critics then share their mathematics-informed interpretations, reactions, and opinions with an audience who might use their critique as a sort of docent-like guide to whatever is the subject at hand.

Art projects are highly generative for teachers of mathematics. They challenge us to orchestrate our classroom in ways that are radically different from the typical math classroom in school. They also promise powerful new ways to use mathematics in life, making mathematics a tool for creative and aesthetic experience.

Further Reading

Appelbaum, P. (2012). Mathematical practice as sculpture of Utopia: Models, ignorance, and the emancipated spectator, *For the Learning of Mathematics*, *32*(2), 14–19.

Bridges Math & Art Organization. (n.d.). https://www.bridgesmathart.org/

Sinclair, N. (2006). *Mathematics and beauty: Aesthetic approaches to teaching children.* Teachers College Press.

CHAPTER 6

Political & Culturally Relevant Pedagogy Lists

- Gloria Ladson-Billings' Six Habits of Highly Effective Teachers
- Celebrating the Mathematical Brilliance of Multilingual Learners
- Levels of Democratic Mathematics
- Post-Colonial Mathematics Levels
- Anti-Racist Mathematics

Gloria Ladson-Billing's Six Habits of Highly Effective Teachers

- Students are intellectual leaders in the classroom,
- Students are apprenticed in a learning community rather than taught in an isolated way,
- A wide range of mathematical literacy is embraced,
- Students' real-life experiences are legitimized as part of the curriculum,

© PETER APPELBAUM, 2024 | DOI:10.1163/9789004688469_007

- Students and teachers challenge the status quo, and
- Teachers see themselves as political beings.

When Gloria Ladson-Billings observed teachers considered highly effective by other teachers, administrators, counselors, and families – that is, teachers truly identified as successful, especially with marginalized populations – she called them *Dreamkeepers* (Ladson-Billings, 1994), and the commonalities among their teaching approaches "culturally relevant pedagogy". When she focused on mathematics teachers and the experience of those learning to teach from experienced, highly effective mentor teachers (Ladson-Billings, 1995), she identified six generalizable ways to describe their teaching. Highly effective teachers do not "teach" in the sense that they are better described as master mathematical thinkers who interact with their students as apprentices. They recognize their students as already mathematicians – OK, perhaps immature and not displaying their mathematical craft in the same way as a very experienced mathematician. Nevertheless, the teachers create more of a studio environment in the classroom, where the apprentices are responsible for running this place of mathematical creation. In such a classroom, student ideas drive the conversation and steer the activity. Student questions and demands for clarification of their own thinking as well as that of others turn them into intellectual leaders, and in this way, Ladson-Billings identifies a similarity with the goals that are discussed in the Alice Pitt Assessment list (See Chapter 2).

Students' ideas and suggestions to others directly influence the direction of mathematical exchange; their contributions become part of the ongoing construction of the mathematical canon within their community. Students who read, write, listen and speak mathematics in such a community are working with a broad conception of mathematical literacy. They employ funds of knowledge (González et al., 2005) in ways that legitimize the mathematics of their everyday lives. The teachers studied by Ladson-Billings guide their students in connecting the mathematics they are learning with past life experiences, dreams they have in their lives, and with ways to make a difference in their community.

You can imagine that the students and teachers in these classrooms are aware something different is happening in their apprenticeship space. The very existence of culturally relevant pedagogy challenges the status quo by questioning most of the basic assumptions of mathematics education. Instead of classrooms where teachers provide instructions for obtaining answers to types of mathematical questions in isolated contexts separate from everyday life, students are designing their own ways of asking and answering questions. They are not allowed to sit back and wait to be taught, as if the magic of the teacher's entertaining lecture will miraculously make them "know stuff". Indeed, the way that the curriculum is unfolding in these spaces can feel threatening to others. The students know this, and they celebrate how much their teacher cares about their learning. They know that the teacher is taking risks, and they know that their teacher is taking these risks for their benefit. They react in kind, loving that they are challenging the status quo and others' low expectations for their engagement and academic achievement. By working in these ways with students, teachers facilitate all members of the classroom community to enact a politics of mathematics education that promises social transformation of the institution of schooling.

How can you use this list to introduce culturally relevant pedagogy in your teaching? Instead of a list of specific techniques, this is a list of guides to figure out how to do this in your own, unique way, much like the dreamkeepers studied by Ladson-Billings. You take each item in the list, and ask yourself, how am I already doing this? How am I not really fully accomplishing this? What would other teachers, administrators, counselors, and family members say about how I accomplish this for each of my students, if they observed my classroom? You ask your students how this is and is not happening, once or twice a month, as part of classroom feedback and student self-assessments. Over time, you use this reflection to make small changes in how you run your classroom. You share your thinking about this with your students, especially the reasoning behind this: These ways of being a highly effective teacher are challenging me and you; if we can make this happen, the excitement of challenging the status quo might just give us a thrill! It can be threatening. At first, the different ways of working here will feel unfair. They will seem wrong. How will we keep our enthusiasm alive? What indicators will we use together to know we are making this happen?

Further Reading

Aguirre, J. M., Mayfield-Ingram, K., & Martin, D. B. (2013). *The impact of identity in K–8 mathematics learning and teaching: Rethinking equity-based practices.* National Council of Teachers of Mathematics.

Parker, Y., Matthews, L., & Jones, S. (2022). *Engaging in culturally relevant math tasks: Fostering hope in the middle and high school classroom.* Corwin.

Celebrating the Mathematical Brilliance of Multilingual Learners

- Position Multilingual Learners as Classroom Leaders
- Enable Increased Participation in Mathematics
- Facilitate Partnerships with Peers
- Use Culturally Relevant Contexts
- Connect with Visuals and Gestures, Sandwich English and Other Languages
- Increase the Reading, Writing, Listening & Speaking in Mathematics
- Engage with Families in Creative Ways

Mathematics teachers who work with multilingual students and students whose home language is not English have the extraordinary opportunity to celebrate their students' abilities to think and communicate in more than one language, to serve as mathematics ambassadors to more than one language community, and to compare and grow from the differences in mathematics concepts when they are articulated in different languages or when they are explored in different cultures. What an exciting and stimulating learning experience for the teacher as well as the students!

Learners as Classroom Leaders

Because the students are multilingual, and because they may be exposed to comparable cultural forms of mathematics when they speak with family members or neighbors, they can share what they have learned in these contexts or critique the standard versions of mathematics typically taught in our schools. Either way, they contribute in ways that are special and are fantastic resources for everyone in the classroom. The trick is not to single out a student as a "representative" of an entire community, but instead to be grateful when they do share, because any sharing of something different, interesting, even similar from other contexts makes classroom discussion possible, and raises the potential for conceptual development. Unfortunately, immigrants and members of minority groups have traditionally been perceived within a deficit culture, which made any of these possibilities virtually impossible. What we seek are ways to *position the learners as classroom leaders*, and one of the best ways to do this is to encourage their participation in general, in much the same way as we should facilitate all students in sharing alternatives and challenging assumptions, whenever they can.

Increasing Student Participation

Incorporating interactive learning activities that include opportunities to practice social and academic language – whether in a home language or in a target language expected by their school environment – supports relationship building with other students while hopefully creating a welcoming environment. Sharing a common mathematics investigation can provide practice in speaking and listening through the mathematics, rather than isolating language learning in an artificial lesson situation specifically about language. Especially helpful in this sense are scaffolds like sentence frames that students can use to communicate their thinking, modeling and repetition of language, and introducing specialized vocabulary as need to explain an idea that they are observed to have constructed through their work. Including your own desire to develop language facility in students' home languages can provide important moments of student participation as well.

Partnerships with Peers

Bilingual and multilingual classroom practices offer plenty of ideas for increasing student participation when language is a possible barrier. We can start by simply including multilingual students instead of passing them by, expecting them to participate, and by positioning them as teachers of their language and perspectives to the other students. Ideally, students should not only be able to explain their own strategies and conjectures, but also understand those of others, asking clarifying questions and listening carefully. Learning across language differences is an effective instructional technique, even if it seems frustrating at first (See "Hannah Arendt's Four Strategies of Unlearning" in Chapter 4). Not relying only on words, but using representations and gestures can scaffold successful communication. As monolingual English speakers learn non-English vocabulary and experience the struggle toward understanding concepts in another language, these can serve them well in the future. And teaching something in a non-mother tongue is one of the best pedagogical approaches to learning that language on top of all of this.

Culturally Relevant Contexts

All students can benefit from problems set in diverse multicultural contexts while doing the research to learn from the elders in their families and communities. This will go a long way toward the students feeling welcome and celebrated (See "Levels of Democratic Mathematics" and "Post-Colonial Mathematics Levels" in Chapter 7). The expanded range of culturally relevant contexts can only serve all students as it provides more possible settings for embedding mathematics, enriching the resources for conceptual connections and examples.

Visuals, Gestures & Sandwiching

Increasing and amplifying our gestures can add visual emphasis to our communication, which serves all students and creates yet one more informal form of representation. We can think explicitly about our gestures, and intentionally increase our use of them. We can also video ourselves and use the video as an assessment of our gestures, to see how students are reacting, to view how big, obvious, or consistent they are, and to consider new possibilities to add to our gesture repertoire. Combining gestures with the *sandwiching* technique is a research-based idea. If our goal is to increase English comprehension, then we would say a phrase in English, repeat the phrase in one of the students' home languages, and then repeat the phrase in English. We can pre-plan key phrases as part of our lesson and unit planning by using online translators or phone apps, and practicing pronunciation by copying the apps speaking option. Not being perfect is part of the technique: we are modeling for our students the effort to try, and we ask them to correct our pronunciation and to help us practice learning their home language in a way that is similar to how we want them to practice English and develop increased fluency in English. If our goal is for ourselves and our students to develop multilingual competencies, then we would sometimes sandwich the phrases in the opposite way, with the English (or another language present in the class) in the middle. This enables everyone to learn the less comfortable language along with the conceptual understanding that is communicated by the outer layers of the sandwich.

Supporting All Forms of Mathematical Literacy

By increasing the uses of all forms of mathematical literacy – reading, writing, listening, speaking and depicting, we are not limiting our students' communication to only speaking, and we are placing each of the literacy modes in relation to the others, so that they are all developed together in support of each other. Ideally, we enable all students to develop skills of each literacy mode in more than one language, so that they can practice communicating across different languages and within different cultures.

Creative Engagement with Families

Finally, celebrating multilingual students' successes in learning mathematics is a collaboration among the school, the class, the students' families, and the extended family of relatives and neighbors. Newsletters translated with online translators can communicate information about what is happening or soon to take place in the classroom. Looking over online translations can help you learn other languages, as well. Meeting with families in local religious centers, community gathering places, and shopping for food in local supermarkets can help you learn more about the everyday life of your students. Asking families to share what they know are the mathematical strengths of your students is powerful information, but more important for this list is to ask the families to share keywords for mathematics that they learned in school, with possible English translations to help you learn this vocabulary that might be new to you. Reaching out to families can help in providing a welcoming environment for multilingual students. Information in their home languages helps to keep families informed. The more you can tap into the families as language resources, the better you can learn about your students' everyday life.

Encouraging families to maintain their home language, and encouraging students to practice telling their families about what they are learning in their home languages, helps students to develop conceptual understanding through home connections, values the families' home cultures and ways of life, and will help students to develop more than surface speaking fluency in their home language as they move through school. Some teachers solicit video from families of mathematics in their lives. What they choose to share, and how they share it, is fascinating, and can be a resource for classroom discussions. Suggestions in email or website updates for simple out-of-school conversations or activities that support concept development taking place in school provide interest ways for families to know that they are supporting their students. For multilingual students, an excellent idea is to encourage out-of-school activities connected to current school topics in the home language. Sharing algorithms the older members of the family learned when they were in school will likely provide contrasting procedures that can be compared and contrasted with standard algorithms. Language differences often mirror conceptual nuances that matter for deeper understanding of the relationships embedded in the procedures.

Engaging multilingual learners means leveraging their energy and curiosity, their diverse experiences, and their creativity as communicators. Multilingual learners can no longer be the few students in the back of the room, whose ideas we'll tap into later when they're able to share them with us in English. Their participation in class discussions and critical thinking processes is an increasingly important element of our shared future. We will need students with the cultural, linguistic, and intellectual flexibility to be able to cross borders, bridge communities, and help solve the global crises we face – whether climate change, ethnic conflicts, or perhaps the next pandemic. (Nordmeyer, 2021)

To provide English language learners with equitable mathematics learning opportunities, researchers suggest that mathematics teachers must develop capacities for and self-efficacy with "linguistically responsive teaching", which involves recognizing, addressing, and capitalizing on students' individual sociocultural, linguistic, cognitive, and academic strengths and needs; specific orientations, such as a view that language, culture, and identity are interconnected; an appreciation for linguistic diversity; and an inclination to advocate for multilingual learners (DePiper et al., 2021). It also involves a set of pedagogical dispositions, such as wanting to learn about English language learners' linguistic and academic backgrounds, probing the language demands of classroom tasks, and scaffolding instruction. If we take the approach advocated in this list, then we are working on language learning for *all* students, not only the English language learners. Such an environment makes the dispositions a moot point, and opens up professional learning over time to include an expanding repertoire of instructional approaches. In other words, thinking about a multilingual classroom, no matter whether there are English language learners, make teaching even more fun and interesting!

Further Reading

Chval, K., Smith, E., Trigos-Carillo, L., & Pinnow, R. (2021). *Multilingual students: Positioning English learners for success, grades K-8.* Corwin and NCTM.

Colorin Colorado (n.d.). *Being bilingual is a superpower: Multilingual tips to support language and literacy at home.* Videos in 8 languages. https://www.colorincolorado.org/videos/being-bilingual-superpower-multilingual-tips-support-language-and-literacy-home

Nordmeyer, J., Boals, T., MacDonald, R., & Westerlund, R. (2021). What does equity really mean for multilingual learners? *ASCD*, *78*(6). https://www.ascd.org/el/articles/what-does-equity-really-mean-for-multilingual-learners

Sales, H. (2022). Indigenous families' rights to translated information from schools. *Colorin Colorado*. https://www.youtube.com/watch?v=cEi_LiJoGmA

Sesame Street. (2021). *Spanish is my superpower.* Sesame Workshop. https://www.youtube.com/watch?v=YXmnUQrHE6s

Levels of Democratic Mathematics

- True-False Ideology
- Right Method
- Appropriateness and Reliability of the Mathematics for its Context
- Question the Appropriateness of Formalizing the Problem for a Solution
- Critical Mathematics Education
- Reflective Thinking as an Evaluative Process

Considering the need for highly effective teachers of mathematics to understand themselves and their work as political, it can be helpful to understand the mathematics we are teaching in terms of the ways that it does and does not support democratic processes (Keitel et al., 1993). When does our mathematics support our students as political actors in the world, helping to make society a better place for themselves and others? There is a way in which the presentation of the mathematics, the style of interaction with it, and the uses to which it is put, together create a "story about what mathematics is". Mathematics as a discipline becomes a kind of tool or technology that structures how we see our world, how we understand "knowledge" as something we can learn and use, and how we use mathematics to relate to others as we build community (Or not!).

A lot of school mathematics can be characterized with the first approach in this list. Definitions, concepts, solutions, algorithms, and so on, can be labeled correct or incorrect, right or wrong, true or false. There is a kind of security to this approach, a simple and satisfying way of knowing the expectations others have for us, that we hold for others, and even what we anticipate for the mathematics we are learning. The second "level" guides teachers and students to ask about the fit of the mathematics to the purposes at hand: are there other algorithms we can use in this situation? Why that one, why not another? What if we step back from the given question and check if we are answering the right one? (A simple example of this: instead of asking for the total cost of a list of items purchased, students might question the choices of the items, the accuracy of the prices described, the health impact of the choices and some alternative totals that are healthier, comparison totals for nearby stores, and so on. Meanwhile, is the total cost the right question to ask, or would these students be better served by finding the percentage of the family's income?) As we move through the levels, we can see that the mathematical skills and concepts remain the same, but our relationship to them changes. They become more relevant to democratic participation. This second level can easily push us into the third, in which we ask about the implications of the mathematics we have been doing. What is the meaning of our answer? Why are we performing this mathematics? How reliable is this particular mathematics for guiding us in making choices, for identifying the right questions to ask and try to answer, for the ways we organize policies in our community?

Questioning the Appropriateness of Formalizing a Problem for a Solution

Level four in this list challenges the use of mathematics itself. In the context at hand, is mathematics the way to go? What other ways might we discuss this situation? Are the statistics collected really the most important kinds of information? Is the graph misleading? What things are hidden by the use of mathematics here? This is not to say that mathematics is wrong, but instead to appreciate the power of mathematics – and also, its limitations. (Another example is useful here: Ole Skovsmose told the story of a Danish middle school classroom. Small groups of students were given the task to create a formula that an office worker could use to decide how much money should be given to each family in the city. They discussed possible relevant numbers, such as total income, number of children and adults, special medical costs, and so on. Of course, different groups came up with different criteria. Once each group produced a formula they thought fair and useable, the teacher prepared special family cases that would demonstrate the failure of the various formulas to achieve the outcomes they had anticipated – such as, some families would get too much, others not enough ... Students tweaked their formulas; the teacher created yet new cases that continued to lead to unfair outcomes. The lesson learned: no formula will be perfect; all will fail to live up to expectations. Now we still have to decide what to do.) A mathematical/technological approach is not always wise "enough".

Critical Mathematics Education

The fifth level of mathematics for democracy has come to be known as "critical mathematics education". Hints of this are part of the previous levels, but within critical mathematics education learners are more careful to ask how particular algorithms and mathematical categories of knowing things affect perceptions of parts of reality. In this way, the role of mathematics in society becomes a component of reflective mathematical knowledge itself. (An example for this level comes from an ethnomathematics perspective. Ethnomathematics challenges the universality of mathematics, recognizing that there are different forms, styles and types of mathematics found around the world, and in different cultures. In particular, school mathematics reflects how the dominance of Western European mathematical history has come to essentially define what we think mathematics "is". The legacies of colonialism have left us with a loss of other histories and traditions, almost entirely erasing them from existence. Entire epistemologies have been obliterated. Indeed, structures of school mathematics, and the universality of the school mathematics curriculum, are consistent with other forms of economic, cultural, geographic, and political inequities among nations and cultures. Some folks directly challenge this, labeling the global structures "coloniality" that perpetuates injustice; they use indigenous mathematics to challenge these structures as part of social and political reform movements. Others look to indigenous mathematics as hope for saving the planet from the devastations wrought by Western mathematics and its assumptions about power and knowledge.)

Reflective Thinking as an Evaluative Process

The sixth and most highly sophisticated level of mathematics for democracy strives to look at the idea of these levels as itself a tool for thinking about knowing, learning, and reflection. At this level, reflective thinking evaluates the other levels: the first two can be thought of as essential mathematical tools for achieving results. In certain situations, mathematics is indeed right or wrong, true or false, and it can be useful to group specific problem types into categories that are solvable in specific, generalizable ways. Levels three and four point to the need for learners to consider the relationship between means and ends. One should always be aware that we can get caught up in the mathematics skills and forget why we are using them. What was that question, again? And, did I answer it? What does that number or shape mean for a decision one would make? Does it make sense? Level 5, the critical mathematics education approach, makes the global impact of the mathematics important. What is the impact of using formal techniques? Does it blind us to inequalities? Does it lead to us ignoring the environmental impact of what we can achieve through our brilliant engineering? Level 6 makes reflective evaluation a tool in itself, a serious "meta-reflection on reflection!"

When teachers and students plan their classroom experiences by making sure that all of these levels are represented in the group's activities, it is more likely that students, and teachers, can be attributed the critical competence that we envision as a more general goal of mathematics education. Students are attributed the critical competence to judge how and where their mathematical ideas can either make an impact on the world, or how the world can make an impact on their further development of ideas.

Further Reading

Appelbaum, P., & Stathopoulou, C. (2016). Mathematics education as a matter of curriculum. In M. A. Peters (Ed.), *Encyclopedia of educational philosophy and theory*. Springer. https://doi.org/10.1007/978-981-287-532-7_515-1

Skovsmose, O. (2023). *Critical mathematics education*. Springer.

Wolfmeyer, M. (2017). *Mathematics education: A critical introduction*. Routledge.

Post-Colonial Mathematics Levels

- Multicultural & Diversity Education
- Critical Ethnomathematics
- Dignity, Recognition & Reconciliation
- Reappropriation Across Bonaventura de Sousa Santos' Abyssal Line

Math teachers always want to spread the joys and opportunities that mathematics provides to *all* students, young and old, in and out of school. The mantra is "mathematics for all" – a phrase that changes meaning and connotation more often than it is uttered by even the same person! As schooling expanded to more and more students through the last century, mathematicians, government agencies, and curriculum developers tried to figure out *what* mathematics should be taught in schools. Are there core skills for everyone? What level of mathematics is appropriate for which students? Is the content most important, or the processes of thinking like a mathematician? More recently, teachers have used the term "mathematics for all" to help them think about pedagogy – what sorts of experiences and processes, what features of the learning environment, will best help all learners to learn the mathematics they need to know? Might there be a need to differentiate instruction to best achieve learning for all students? Are some ways of learning more pleasant, so they will help students want to be there to learn? These questions shift the focus from the mathematics concepts and skills to attributes of the teacher and students: do some learners bring different prior experiences to school, which make it easier or more difficult for them to learn in the way that the teacher has orchestrated? Does the teacher have the intercultural communication skills and awareness about diversity necessary to make adjustments to meet the needs of the learners in their classroom? Does "mathematics for all" mean equal opportunities to learn mathematics relevant to a joyful life, or equal outcomes of mastery on a set of core skills that every citizen needs for participation in society? There are of course no "right answers" to any of these questions. Yet, once we have asked them, they cannot be ignored. We need to decide how we will honor the diversity of perspectives on mathematics, variety of cultural backgrounds of teachers and students, and possible instructional choices that are, can, or should be included in our design and implementation of mathematics learning environments.

The 1980s and 90s spawned "Multicultural Mathematics Education" and an increased awareness by mathematics educators of "Ethnomathematics". Multicultural education begins, not with what content should be taught, but instead with what cultural ways of being mathematical are brought into the classroom by the teacher and students (See Alan Bishop's Six Mathematical Activities, Tamsin Meaney's List of Bishop's Activities, and David Kirshner's Three Types of Activity in Chapter 1). Beyond that, there are numerous ways to consider "being multicultural" (Appelbaum, 2002). First, if we consider the sorts of formal and informal learning experiences that our students have had prior to entering our classroom – with family, in their neighborhood, in other schools, perhaps in other countries or different regions of the country, social and cultural expectations about what learning looks like, how one pays attention, what adults assume about children, and more, all matter for how lessons will unfold. What do we know about our students, their interests, their knowledge of the world? What do they know and what can they do that we can't? – that we never would think for someone at that age? What alternatives to our standard "basic skills" and standard algorithms do they find routine, and which routine facts and vocabulary that we take for granted are peculiar or new to them? Taking cultural difference into account means that we can tailor our lessons and pedagogical style to best match our students. Recognizing the diversity among our students means planning and assessment should take this into account as well. In its most positive form, this sort of multiculturalism can be called "Culturally Responsive Mathematics Education" (Greer et al., 2009). In its least positive version, multiculturalism can devolve into "teaching the culturally different" with a "deficit mentality" – rather than exploiting the rich funds of knowledge that students and their home cultures can contribute to the classroom community, this worst case version of mathematics teaching treats learners as requiring remediation to be brought into the mainstream, and denies their special and unique talents and perspectives the opportunity to be recognized and appreciated. More recently, this awareness has moved into mathematics for special education as well. Even teachers who appreciate diverse cultures sometimes need help in overcoming stereotypes about learners with disabilities so that they can work with disabled students in ways that dignify and celebrate their inherent capacities to think and do (Tan et al., 2019).

But that is just a start on multiculturalism! Once we think about how our students know different things from us, and bring different skills than we might have assumed, we can also wonder why some of these things are not already in our curriculum. This is where *Ethnomathematics* is increasingly important: culturally different mathematics means that our curriculum is grounded in one or only *some* cultures. Others are absent. Are there mathematical concepts or skills valued in our students' home cultures that are currently ignored by our curriculum? Are there social contexts or problem-solving situations in their home cultures that are more relevant than the ones used in our text materials? Are there mathematician role models or historically significant mathematical developments from around the world that would be a more diverse set of motivational heroes than the ones featured in our textbooks or that we tell stories about? This "next level" of multiculturalism asks us to revise and redesign our curricular content to meet the needs of our students.

Multicultural & Diversity Education

The first item in this list is not "multicultural mathematics education" but *Multicultural & Diversity Education*. Why? The *Diversity Education* part of this relates to criticisms of multiculturalism that seeped into mathematics education in the late 1990s and early 2000s. One of these is that multiculturalism seems good and feels good but does not really change anything in the world beyond making some of the curriculum a little different. Maybe interpersonal relationships are better once we think about why some student behaviors strike us as odd or inappropriate, which is certainly important, but there are ways that tweaking the curriculum to be more multicultural actually works against itself. For example, adding a few stories about non-European mathematicians, or using the African game of mancala to model arithmetic in elementary mathematics, or to model functional relationships or to investigate probability distributions in more advanced mathematics, does not change the fundamental reality that the mathematics we are teaching has its origins in Europe and Western cultures. Mancala is not interesting because we can see mathematics in it. It is interesting in its cultures of origin for many reasons. And there are other, more "mathematical" activities in those cultures of origin that are not featured because they don't look like the school mathematics – shouldn't *these* become part of the school curriculum? This is yet another "next level" moving us toward *Critical Ethnomathematics*, a form of *Post-Colonial Mathematics Education*.

Critical Ethnomathematics

Critical Ethnomathematics understands ethnomathematics is only possible because of historical circumstances (Appelbaum & Stathopoulou, 2023). School mathematics looks like traditional European and Western mathematics because it is. Colonialism imposed many things upon other parts of the world, including school structures and school curriculum. And these impositions are still with us today. The institutions and geopolitical economic, political and cultural systems that preserve unequal balance between former colonial powers and their colonies are together understood with the term "coloniality". Critical Ethnomathematics introduces a political, social justice orientation to multicultural and diversity education: How does the situation matter? Coloniality erases some forms of mathematical activity from even being known. Examples:

fractal thinking that has been present across various African cultures way before fractals were even a possibility for Western mathematics; aperiodic tiling patterns that cover a plane, only "discovered" recently by professional mathematicians, but known to designers of Islamic mosque walls for centuries; non-Euclidean geometry of space and time, still a challenge to many students of mathematics and physics, but always common to everyday life perspectives in Navajo, Metis, Aboriginal and other indigenous communities world-wide. These topics in mathematics are supposedly too advanced for K-12 schools, yet are only thought of as advanced from a perspective tinted by coloniality.

Dignity, Recognition & Reconciliation

... is a post-colonial effort to live in our post-colonial world, the legacies of colonialism and its erasure of indigenous knowledge, and the associated imbalances. Indigenous practices may not look like mathematics to a "good student" of what is typically taught in schools. What can we do as teachers of mathematics? We need to dignify local practices of Problem Solving and Posing, and recognize everyday life practices that incorporate Bishop's Six Activities as truly mathematical. Reconciliation brings about a mutually respectful valuing of all aesthetic, utilitarian, and culturally expressive forms of game, structure, pattern, design, demonstration, explanation, exploration, location, and so on. How do we do this? We can start by using "culture" in an empowering way, rather than as a form of coloniality. What I mean is, culture is a concept from Anthropology, another example of coloniality – anthropology was born in European attempts to study colonized people as "others", and in this way makes the others into objects of study, rather than as human beings of equal stature. Instead of "studying them", (who is them???) we all work together to avoid seeing the officially important knowledge and ways of doing things as "the culture" (in the way that "being cultured" means liking those things of privilege like fine wine and fancy museum art). When students are taught standard algorithms and required to show their work, does this devalue the algorithms their parents use, which cannot be "shown on paper" in the same way? Better is to study all algorithms for how they are more useful than the others, more beautiful in some ways, easier to follow, and so on.

The next step is to accept that existing curriculum materials are tools of coloniality. They are not going to give you packaged lessons that are examples of "reconciliation". Teachers pursuing reconciliation meet with elders in the community for guidance. If it seems overwhelming or beyond the scope of a job description, then teachers work together to make meetings with elders to become integrated into the school calendar. They learn about what matters to their students and their families. They use mathematics to support community activities, rather than community activities to support a provided curriculum. These teachers learn mathematics themselves through participation in local traditions, crafts, political organizing, and family milestones. Aikenhead (2017) urges us to enable students to simultaneously master and critique mathematical ways of knowing without devaluing or setting aside their culture's ways of knowing, doing, living and being in the process, and use mathematics to strengthen their own cultural self-identities, and declares, "Just as we treat Indigenous mathematizing as cultural practice, we need to teach Euro-American mathematics as the Euro-American cultural practice it is" (p. 171).

We can learn from the history of mathematics in Japan. Japanese culture historically used Wasan (和算) as something similar to typical school mathematics. Suddenly forced in 1868 by American navy gunboats to join the economic globalization movement, Japanese people officially introduced a mathematical system called "Yousan" (洋算 – foreign over the ocean calculation"). Japanese scholars realized that Yousan was a cultural creation that carried with it ideologies, values, and perspectives foreign to those of Japan's. For example, the Yousan counting system uses the same names for numbers no matter what is being counted ("Mathematics is the art of giving the same name to different things" – Henri Poincaré, French mathematician, early 20th century, cited in Verhulst, 2012). On the other hand, the Wasan counting system, which continues today to some degree, has different number names depending on the group of objects being counted -- people, ages, kitchen cupfuls of ingredients, long skinny things, thin flat things, clothes, etc., all of which subtly relate to an endemic Japanese worldview. If one version of counting is called "mathematics" it diminishes the other. Surely a facility with both, along with a critical perspective on each in terms of what it accomplishes as well as ignores, would be the most empowering.

Reappropriation across Bonaventura de Sousa Santos' Abyssal Line

Coloniality establishes some versions of mathematics as "the mathematics" and either erases or diminishes others. Portuguese sociologist and esteemed international scholar Bonaventura de Sousa Santos (2007) uses the metaphor of an "abyssal line", a chasm between one culture and another, to describe how one side of this chasm establishes through everyday life the ways of thinking and learning that are valued as normal, while erasing other forms of knowledge and learning from the world. Reappropriation across the abyssal line is a way to respond mathematically. While some might desire a reclamation and celebration of local cultural traditions that is independent of coloniality, others seek "two-eyed seeing" (Hatcher et al., 2009). "Two-eyed seeing" values the strengths of mathematizing with local traditions *and* Western (school) mathematics, so that the students can view their world through two simultaneous lens. Using bits of either knowledge system, or choosing one over the other, in order to solve a problem or make sense out of an issue, students would bridge worlds and realities, translating between them, and contributing to both. Rather than dismissing traditional school mathematics, these students would want to master it as one of several possible systems of interpreting and contributing to their communities. They might study for inspiration, Ida B. Wells' (1894) *A Red Record* tabulation of lynching statistics and their causes, and W.E.B. Du Bois' (1900) Paris Exhibition of statistics on Black Americans' successes since abolition despite the challenges of reconstruction.

Further Reading

Aikenhead, G. (2017). *School mathematics for reconciliation: From a 19th century to a 21st century curriculum.* https://education.usask.ca/documents/profiles/aikenhead/School-Mathematics-for-Reconciliation-vB13.pdf

Appelbaum, P. (2002). *Multicultural and diversity education: A reference handbook.* ABC-CLIO.

Appelbaum, P., & Stathopoulou, C. (2023). History of ethnomathematics: Recent developments. In S. Romero Sanchez, A. Serradó Bayés, P. Appelbaum, & G. Aldon (Eds.), *The role of the history of mathematics in the teaching/learning process* (pp. 189–207). Springer.

de Sousa Santos, B. (2007). Beyond abyssal thinking: From global lines to ecologies of knowledges. *Review (Fernand Braudel Center)*, 45–89.

Greer, B., Mukhopadhyay, S., Powell, A., & Nelson-barber, S. (Eds.). (2009). *Culturally responsive mathematics education.* Taylor & Francis.

Hatcher, A., Bartlett, C., Marshall, A., & Marshall, M. (2009). Two-eyed seeing in the classroom environment: Concepts, approaches, and challenges. *Canadian Journal of Science, Mathematics and Technology Education*, 9, 141–153.

Tan, P., Padilla, A., Mason, E., & Sheldon, J. (2019). *Humanizing disability in mathematics education: Forging new paths.* NCTM.

Anti-Racist Mathematics

- Use Mathematics for Anti-Racism
- Use Anti-Racist Ideas for Mathematics
- Counter the Backlash to Anti-Racism

As the American philosopher, trans-disciplinary scholar and political activist Angela Davis once said, "In a racist society, it is not enough to be non-racist, we must be anti-racist". This list helps us to think about how to do this as mathematics teachers. Given the legacies of racism world-wide, and especially in countries that owe much of their historical successes to slavery, everyone must do whatever they can to dismantle racism and its never-ending traces. Whether we realize it or not, racism affects everyone, and makes its impact on everyone's daily life. Yet, critics of anti-racism discussions and attempt to translate anti-racism into policies and practices often say things like, "Let's just do the math! Quit this 'woke' garbage!" So how do we make sense of this as mathematics teachers?

Use Mathematics for Anti-Racism

One approach is to use mathematics for anti-racist purposes. We can collect ideas for anti-racist examples that use the mathematics we need to teach. In this way, we provide students with contexts about how mathematics has been used to perpetuate racism, and also give them an opportunity to apply mathematics in real-life scenarios that encourage equity. Racial profiling of police stopping motorists is an example: If we first create a sample of cubes with different colors to represent the racial and ethnic distribution of our local city or region, and pick cubes out randomly, then data collection models a truly 'color-blind' practice. This can be compared with actual data while practicing conversions of fractions and decimals, probability distributions and representative sample questions. Using a 2-liter soda bottle to estimate how many would be needed to fill a bathtub or a pot of water for boiling spaghetti can be the start of a lesson during which students first estimate how much water they or their family use in one week, and then extend that to a discussion of First Nations and Native American communities water advisories for boiling all water, and what this means for their everyday life. Using mathematics as a tool for studying power and privilege has many possibilities, and there are more and more websites with sample lessons that do this.

Another approach is to think about mathematics itself as a tool of racism, and how a different perspective on mathematics can lead to different kinds of mathematics (See "Post-Colonial Mathematics Levels", previously in this chapter). Students might study the ways that their school or school district collects data by race and ethnicity, and what this means for school policies. Do they agree with them as fair and equitable? The history of mathematics used to label people by IQ and other categories can be the context for realizing how statistics and the scientific method owes much of its development to racist purposes. As we re-think our mathematics classroom, we can work with colleagues and our students to discuss the history and present practices of mathematics outside of Europe, North America and other English-speaking, former colonies of the

United Kingdom, and the important contributions of this mathematical knowledge to civilization. We can make plans for how to identify and avoid further racial stereotyping when forming and communicating expectations of our students' perspectives on mathematics, achievement in mathematics and choices about continued study of mathematics. We can work together to challenge stereotypes and cultural biases in our classroom materials, textbooks, topics, and assessment tools.

Use Anti-Racist Ideas for Mathematics

Cintron et al. (2020) have assembled a fantastic tool for dismantling racist mathematics education month-by-month. They identify seemingly okay words and associated ideas that perpetuate white supremacy, and outline exercises that can be used to identifying their sly sneakiness in our teaching, as well as strategies for avoiding them: Perfectionism, Sense of Urgency, Defensiveness, Quantity Over Quality, Worship of the Written Word, Paternalism, Either/Or Thinking, Power Hoarding, Fear of Open Conflict, Individualism, Only One Right Way, Progress is Bigger & More, Objectivity, and Right to Comfort.

Being "anti-racist" requires us to fight against racism. Racism takes a number of forms and they each work to support the others to subtly reinforce racist ideas "behind to scenes" without us necessarily even being aware that this is happening. Individual racism refers to beliefs, attitudes, and actions of individuals that support or perpetuate racism in conscious and unconscious ways. Common cultural narratives about racism typically focus on individual racism and fail to recognize systemic racism. Examples include a belief in the superiority of white people, not hiring a person of color because "something just doesn't feel right", or telling a racist joke. Interpersonal racism occurs between individuals. These are public expressions of racism, often involving slurs, biases, or hateful words or actions. Institutional racism occurs in an institution or organization. Discriminatory treatment, and unfair policy, or biased practices based on race that lead to inequitable outcomes and extend considerably beyond prejudice almost never mention any specific racial group, yet the intent is to create advantages, for example, a school system where students of color are more frequently distributed into the most crowded classroom or least comfortable workspaces with underfunding. Structural racism is the overarching collection of these together, which end up being a system nobody is officially responsible for, yet which spreads and maintains racial biases across the institutions and the broader society. Stereotypes of people of color as criminals, or students of color as needing remediation in mathematics, in actual schools or in films and television, are examples of such structural racism.

Countering the Backlash to Anti-Racism

Unfortunately, there has been a recent backlash against anti-racism in all of its forms, and mathematics teaching is not alone in this backlash. Critics claim that anti-racist mathematics takes relativism and postmodernism to extremes, and fall back on beliefs that mathematics and science, especially hard science fields such as physics, chemistry, and biology, accurately reflect universal knowledge, as opposed to being cultural interpretations or social constructions. For example, some critics say that Newton's laws of gravity are neutral formulations that apply regardless of cultural, regional or ideological context. With this view, critics suggest that hard science is different from ideology because it's based on the scientific method, essentially an extremely cautious means of building a supportable, evidenced understanding of our world. They ignore how and when someone might apply those laws of gravity, the contexts in which one would consider them relevant or need them, and so on. Critics also ignore the role of Middle Eastern, Arab, Persian, Chinese, and Indian mathematicians and astronomers from the history of traditional, mostly 'Western' mathematics. The most visible acknowledgement of this history is the use of Hindu-Arabic numerals. Indian mathematicians are also widely recognized as first developing the concept of the zero, and Mayans in Central America also pre-dated European uses of zero. Critics of anti-racist teaching ignore the successes of culturally relevant pedagogy (See "Gloria Ladson Billings' Six Habits of Highly Effective Teachers", previously in this Chapter) and claim that changes in the curriculum or modifications of instructional practices lead to lower standardized test scores, without questioning the validity or usefulness of the test scores in the first place.

How do we *counter the backlash to anti-racism*? We can use actual research-based evidence and professional recommendations from such organizations as the National Council of Teachers of Mathematics and the American Mathematical Association to point out that anti-racist mathematics instruction is yet to actually be fully implemented pretty much anywhere, and that test scores, for example, cannot be blamed on this. We can try to educate critics about the actual history of mathematics, about ethnomathematics and post-colonial issues in mathematics education. However, education is not necessarily successful when the critics are mainly attacking anti-racism because of ideological and political intentions. Instead, we have to consider organizing tactics, rallying the support of colleagues who know about best practices, communicating with families about the needs of their children, and actually 'armoring ourselves' with the evidence that supports what we know about anti-racist mathematics education. We need to recognize that equity in mathematics education is a political act, that teaching is political even when we are not thinking of it that way.

Further Reading

Aguirre, J. M., Herbel-Eisenmann, B., Celedón-Pattichis, S., Civil, M., Stephan, M., Pape, S., & Clements, D. (2017). Equity within mathematics education research as a political act: Moving from choice to intentional collective professional responsibility. *Journal for Research in Mathematics Education, 48*(2), 124–147.

Cintron, S., Wadlington, D., & ChenFeng, A. (2020). *Dismantling racism in mathematics instruction: Exercises for educators to reflect on their own biases to transform their instructional practice.* Stride. https://equitablemath.org/wp-content/uploads/sites/2/2020/11/1_STRIDE1.pdf

Gonzalez, L. (2021, December 15). Critical race theory in the mathematics classroom. *The Activist History Review.* https://activisthistory.com/2021/12/15/critical-race-theory-in-the-mathematics-classroom/

Gutiérrez, R. (2018). When mathematics teacher educators come under attack. *Mathematics Teacher Educator, 6*(2), 68–74.

Malcolm, X. (1965). What did Malcolm X think of white liberals? *CBS Front Page.* https://www.youtube.com/watch?v=d7ZZPeZc6ps

Martin, D. B. (2019). Equity, inclusion, and antiblackness in mathematics education. *Race Ethnicity and Education, 22*(4), 459–478.

CHAPTER 7

PlayBook Lists

- Problem Solving & Posing Lists
  - Georg Polya's *How to Solve It*
  - John Mason, Kaye Stacey, & Leone Burton's *Thinking Mathematically*
  - Stephen I. Brown & Marion Walters's Art of Problem Posing
- Meaningful Communication Lists
  - Concept-Based Lesson Genres
  - Promoting a Thinking Classroom
  - Five Dimensions of "TRU"-ly Powerful Classrooms
  - Peter's Practical Critical Thinking Strategies
  - Five Days of Better Question Asking
  - Worksheet Options
- Storytelling & Literature Lists
  - Daniels-Type Literature Circle Roles
  - Storytelling Techniques for Math Teachers
  - Three-Act Mathematics

© PETER APPELBAUM, 2024 | DOI:10.1163/9789004688469_008

Problem Solving & Posing Lists

Georg Pólya's How to Solve It

- Understand the Problem
- Make a Plan
- Carry Out Your Plan
- Look Back (& Reflect)

The first of our problem solving and posing lists comes from the classic approach attributed to the mathematician Georg Pólya, who taught at the University of California at Berkeley in the 1900s (Pólya, 1945). Frustrated with his students' inability to solve problems that he presented, Pólya analyzed how he did it himself, and observed his colleagues at work: how does a mathematician solve a problem? He realized that a mathematician asks themselves questions that keep them on track. He found a lot of success in modeling these questions for his students, for example by thinking aloud, asking himself questions and trying to answer his questions himself in front of his students. Over time, he was pleased that students could use his questions to succeed in solving problems, and, even more happily, they would eventually internalize the approach, asking themselves these questions as a guide through the problem-solving process. To start, certain questions can help you to *understand the problem:*

Understand the Problem

- Have I ever seen a problem like this before?
- What do I know in this problem, and what do I not know?
- Are there any possible variables in this problem?
- Can I draw a picture?
- Are there any words or relationships in this problem that I don't understand, so that I should look them up?

It will of course be the case that you cannot answer every one of those questions. Nevertheless, if you can answer one of them, it can get you started on making a plan. If you can answer more than one of those questions, you can try out more than one initial plan to see which one gets you going on a good path toward a solution (See Table 1).

Carry Out Your Plan

If all goes well, the plan flows nicely from the way you understood it. Sometimes, you are stuck, even after using the ideas in Table 2. If that's the case, try another way to understand the problem!

TABLE 1 Make a plan

If *yes* to understanding that question ...	Then a plan could begin with ...
Have I ever seen a problem like this before?	Try out that same method in this problem
What do I know in this problem, and what do I not know?	Make a chart of knowns and unknowns, and describe to yourself how the unknowns are related to the knowns. Can what you know help you figure out what you do not know?
Are there any possible variables in this problem?	Create an equation with the variables, based on the relationships described in the problem.
Can I draw a picture?	Draw the picture, label things, use the picture to make sense of what is going on in the problem.
Are there any words or relationships in this problem that I don't understand, so that I should look them up?	Find out what these things mean and start over.

TABLE 2 Carrying out your plan

If you began with ...	... then carry out your plan by ...
Using a method from a previous problem	Obtain a result with that method, and think about what it means.
Making a chart of knowns and unknowns, and describing how the unknowns are related to the knowns. (Using what you know to help you figure out what you do not know)	Use the chart to go back to step one: Are you reminded of a similar problem seen before? Are the knowns and unknowns possible variables? Can you draw a picture now? Create an Equation?
Creating an equation with variables, based on the relationships described in the problem.	Solve your equation and think about what that result means.
Drawing a picture, labeling things, and using the picture to make sense of what is going on in the problem.	Add things to your picture, and describe to yourself and maybe in an equation, how what you have added to the picture is related to other parts of the picture.

Look Back

When you pause or reach a result of some kind, it is important to proceed to this fourth step in the process, and ask yourself:

- Does my answer make sense?
- What is the meaning or my answer? What are the implications of what I have done or found?
- Looking over what I did, might there have been an easier or otherwise better way to do all of that?
- What new questions do I have now, at this point, prompted by reflecting on the experience of this work?

Of course, the steps may not go one after the other. It is likely that you will bounce around them, revisiting them as you try them out. The important thing is that the questions will get you unstuck and refocused on doing something – they each provoke something to think about or write down, and that gets the flow re-started. Note too that the last questions start the process all over again, in a wonderful, never-ending cycle: those new questions turn a problem into an exciting adventure, an investigation beyond the initial statement of a problem.

Further Reading

Pólya, G. (1945). *How to solve it.* Princeton University Press.

Ray-Riek, M. (2013). *Powerful problem solving: Activities for sense making with the mathematical practices.* Heinemann.

John Mason, Kaye Stacey & Leone Burton's Thinking Mathematically

- Specialize – Try Special Cases
- Collect & Organize Your Cases by Categories or Patterns
- Generalize Your Categories or Patterns into Conjectures
- Test Your Conjecture with Artfully Chosen New Cases
- Convince Yourself, Convince a Friend, Convince a Skeptic

Our second problem solving and posing approach shares a lot in common with observational science, something like field biology or astronomy, where the scientist cannot create a test that generates data about an intervention-based hypothesis, but instead must use whatever data they can collect through diligence and creative patience (Mason, 1982). If your problem has to do with numbers, you start by trying special numbers to see what happens – 0, 1, 2,

10, 100? How about odd and even numbers? Prime numbers? Triangular numbers? What else? If your problem has to do with shapes, you try special shapes – circles, squares, rhombi, hexagons? Two- and three- dimensional shapes? Big areas and smaller areas?

The key to success with this approach is to organize the data you are collecting with your special cases. If you try the odd numbers up to 100, in order, you might make a table of your results. If you try powers of ten, in order, you might look for a pattern in a chart. Ordered records of shapes can be helpful. The trick is to look for categories with which you can group results, patterns that you can describe, if-then relationships that you are starting to see. Once you get to this point, you are ready to call what you describe a *conjecture*, and then you can test your conjecture. Does it hold for more cases? Can you challenge the conjecture with outliers that actually do *not work?* That would be awesome, because you can refine your generalization to say that the conjecture only works in a narrow range of categories or situations. Once you are sure that you have a really good handle on this, you can describe your conjecture more carefully, and write out or record a spoken explanation to yourself for why you know that this conjecture makes sense.

That first explanation is one that makes sense to you yourself. Now you proceed to prepare an explanation for someone else, a critical friend who is inclined to believe you and who understands a great deal of the mathematics that you have been working with, so that they can help you to think about how to prepare an even more formal explanation for someone less ready to just believe whatever you say. This third explanation is one that you might need to re-write a few times, with feedback from different friends. When it is ready for a skeptical audience, you can try it out and see how it goes.

Further Reading

Mason, J., Burton, L., & Stacey, K. (1982). *Thinking mathematically.* Pearson.

Wells, D. (1995). *You are a mathematician: A wise and witty introduction to the joy of numbers.* Gardner's Books.

Stephen I. Brown & Marion Walter's Art of Problem Posing

- Find a Problem Solved or a Common Ordinary Relationship
- List Attributes of that Problem or Relationship
- Ask What-if-Not Aout One or More Attributes
- Work on the Newly-Posed Problem, or Look for a Family of Problems that Share the What-if-Not Resemblance.

Problem Posing

Possibly far more important than problem *solving* is problem *posing* (Brown & Walter, 2005). If you think about it, pretty much every "problem" given to students to solve in school is already solved by tons of people over many millennia. Why are we solving them again? Okay, I get it: it's good practice and helps students learn. However, is this the only or the best way to learn? Mathematicians don't always solve problems that others give them. They spend a lot of their time posing their own problems to solve. Indeed, it is something of an ethical concern that students are implicitly taught in school that mathematics is all about solving problems that others pose for you. The "lesson learned" is that you should be passive and accept that others' questions are more important than your own. Your job is not to ask things, but to carry out the tasks posed by others.

As Stephen I. Brown was so fond of asking, even if our goal in mathematics is to help students to become really good problem solvers, does this mean that solving problems is the only way to help them get there, or indeed, the best way? There are plenty of other things to do with problems besides solve them. For example: Why not compare problems to identify what is common or different among them, and generalize what you are learning beyond a specific problem or a type of problem? A class could discuss a collection of problems and put them into categories. They might even put them into categories more than one way, and discuss how the categories highlight attributes that are mathematically important to them.

Why not take a problem and consider whether it is a good one? The most important one to think about right now? Who asked this question, and do we think they are asking the right question? Besides categorizing problems, we might critique textbook problems and change them into the problems that we believe should have been in the book. Some teachers encourage their students to write to the book authors or publishers with their suggestions. Others know that authors and publishers are busy and may not respond, so they have their students re-write chapters of their textbooks for future students to use.

Start with an Already Solved Problem, or an Ordinary Relationship

Brown and Walter (2001) wrote a book on this very topic. One approach is to start with a problem you have solved, or an ordinary mathematical relationship, and brainstorm any and all questions that pop into your mind when you think about it. Which of these things can be thought about mathematically, and which not? (For example, changing the color of tiles in a word problem about how many tiles are needed is not usually a mathematics question. But changing the dimensions of the tiles, their shape, or a pattern with them, is more likely to be a mathematical situation.) Now you have questions that can be explored, *voilà*!

Attribute Listing

Here is a more formal approach: First, take an existing problem or mathematical statement about a relationship, and list everything about it – this is called "attribute listing". Then create a new problem by changing any one attribute. For example: "Kavon and their family are buying tickets for the carnival. There are 3 adults – Kavon's dad, grandmom, and aunt. There are 4 children under 15: Kavon, his 2 sisters, Maria-Louise and Anita; and one of Anita's friends Zelda. Adult tickets are $17.50, and tickets for children under 15 cost $9.00 each. How much will tickets cost for Kavon and his family to go to the carnival?" First, we can list attributes. Here is a starter list.

- A family is buying carnival tickets
- We want to know how much it will cost for the family to buy these tickets
- The family has 3 adults
- There are 4 children under 15
- One of the children is a friend of the family
- Adult tickets cost $17.50 each
- Children under 15 can buy a ticket for $9.00 each

What-If-Not?

Now we can ask, "What if not?" about any one attribute, and change it to pose a new problem:

- Maybe the family is considering the carnival as one option for the day. For this much money, what else might they do?
- Is this much money reasonable for a fun outing, considering a typical family budget? How much of a typical family's monthly budget might this be? How does this compare with a trip to the supermarket?
- What if dad stayed home, or auntie? What if more adults joined them? What if other families went together – are there group discounts at this carnival?
- Surely there could easily be fewer or more children as well …
- Why only one friend of the family? Wouldn't it be more fun to celebrate Kavon's birthday next month with such a trip, and if so, who might come?
- Is this a real carnival ticket price? Does it make sense? What if an adult ticket cost a different amount? Let's consider several different possible ticket prices for adults, and how that would change the total cost. Then we could go back to other questions here and consider whether the comparison with a monthly budget or other options would change.

- Similar to adult ticket price changes, what if children under 15 cost a different amount? What are their ages, actually, and what if the tickets are sold with a different set of prices, such as: free under 6, $6.00 6–10 years old, $10.00 11–16, $20.00 adults, $5.00 seniors?

... and now we have only begun on our journey of new questions ...

The really interesting thing that happens when we start asking these new questions is that we sort of explode the possibilities of what the new question might be. For example, when we propose new ticket prices, we could really imagine any prices for any number of ticket categories. Yet, when we do, the process of determining the total for Kavon's family is pretty much the same. We multiply or add ticket prices for each of the same ticket category, and find a total sum ... (oh my ... do we need to add taxes and parking fees? – that's another question with a different process for determining the total!). What we get once we try out a few different sets of ticket prices are a bunch of special cases. This is a lot like the Specializing and Generalizing in the Thinking Mathematically playbook! (previously in this chapter). By collecting enough of these special cases, we can describe some general patterns and make important conclusions, such as:

- Ticket prices above $100 for a carnival make the family ticket total far too high for most of the families in our school.
- More than 3 or 4 ticket price categories just gets too complicated. It's probably not worth it for a carnival to break it down into more age levels, or special prices for military veterans, teachers, or adults in wheelchairs (or maybe it is? – something for a class to consider!).
- We can create a general algorithm for a family to follow: [(number of kids under free age) × 0] + [(number of kids 6–10 years old) × (price for this range0] + [(number of kids 11–16) × (price for this range)] + [(number of non-seniors) × (adult price)] + [(number of seniors) × (senior price)] + (parking or other transportation costs) + (budget for snacks and drinks) = Total Outing Cost.
- Now we can explore how any one of the things in the general algorithm effects the others. The number of free age kids does not affect ticket price totals, but it might matter for transportation, food or drinks! The higher ticket prices make more of a difference in the total than the other ticket prices, but the adults might not need so many snacks. If we start with a fixed maximum budget for the total outing cost, it might change how any of the numbers of people or costs on the other side of the equation. If we experiment with holding some of these "variables" constant, and change the

others, we get families of problems with characteristics. In these families of problems, different variables matter more than others do, in different ways. – you can see how discussing these things is more mathematically rich than simply solving any one of the potential problems that might be asked for a particular family going to a specific carnival.

Further Reading

Brown, S. I. (2001). *Reconstructing school mathematics: Problems with problems and the real world.* Peter Lang.

Brown, S. I., & Walter, M. (2005). *The art of problem posing.* Lawrence Erlbaum.

Meaningful Communication Lists

Concept-Based Lesson Genres

– Classifying Mathematical Objects
– Interpreting Multiple Representations
– Evaluating Mathematical Statements
– Exploring Problem Structures

Although mathematics teachers know in theory that they should facilitate conceptual development, and that procedural skills and application of algorithms is built upon conceptual understanding, it is a never-ending quest to figure out how to actually make this happen. Concept-based lessons come to the rescue! They help us focus on *what the students are doing*. If they are mostly doing one of the things in this list, then they are more likely to be experiencing conceptual development than not.

Classifying Mathematical Objects

As discussed in "Art Projects for Math Classes", "Susan Sontag's Alternatives to Representation", "Peter's Practical Critical Thinking Strategies", "What Every Mathematics Classroom Never Knew It Needed", and "A Short List of Mathematical Games You May Have Overlooked", anything can become a 'mathematical object' if we decide to 'mathematize' with it. More 'mathy' math objects are things like numbers, shapes, functions, arrays, graphs, and so on. Concept-based lessons can and should at some point involve students in carefully examining collections of mathematical objects and *classifying* them according to attributes (See "Problem Solving and Posing Playbooks"). When students select an object, discriminate between that object and other similar objects (what is the same and what is different?), and create and use categories to build definitions, they are tapping into the emerging mathematical power behind different mathematical terms and symbols. They are turning the special words and symbols into shorthand for the processes through which they were developed.

Interpreting Multiple Representations

When students share different representations, on the other hand, so that they can be *interpreted*, that is, so that they can be shared, compared and grouped, it helps them to construct meanings and connections among the underlying concepts that these representations are in fact 'representing'. Mathematical concepts do have many representations. When students translate across them, from any one of words, diagrams, algebraic symbols, tables, and graphs to one or more of the others, they need to go beyond the surface manipulation of these representation, past copying what a teacher demonstrates or another student performs and gets praise for, toward using these representations themselves to communicate what they are thinking and what they are wondering about. (See "Hannah Arendt's Four Strategies of Unlearning" in Chapter 4.)

Evaluating Mathematical Statements

A very different sort of activity that also pushes conceptual development presents a collection of mathematical statements or generalizations. Students are asked to decide on their validity, and to give explanations for their decisions. This is *evaluating mathematical statements,* and deserves practice in itself, separable from practice in applying algorithms or explaining concepts. Sometimes, the statements may summarize recent student misconceptions, or misunderstandings that a teacher anticipates based on past experience. (For example: "The square root of a number is smaller than the number" (Shell Centre, 2012, p. 7). *If* a teacher models exemplary explanations to include examples and counterexamples that support or refute the statements, then students can practice this approach to clear mathematical communication. Further apprenticeship in mathematical thinking and the development of generalizations involves coming up with conditions that limit the statement so that it can be valid within those conditions, or to otherwise revise a statement so that it doesn't need special conditions and is 'always true'.

Exploring Problem Structures

Yet another approach asks students to *explore problem structures* (See "Problem Solving and Posing Playbooks"). In classrooms where students are used to routinely organizing and classifying problems and questions into types by their attributes, they can be encouraged to devise their own problems and questions that have similar structures. They should try to create problems that are both challenging and that they know they can solve correctly, as well as questions for which they have no idea where or how to begin. Students might first solve their own problems and then challenge other students to solve them. Or they might bring a confusing question to an open mathematical salon discussion. In either situation, they can offer support and act as 'teachers' when the problem solver becomes stuck, ask clarifying questions, and pursue community-building processes through mathematics. Creating and solving problems may also be used to illustrate doing and undoing processes in mathematics. For example, one student might draw a circle and calculate its area. This student is then asked to pass the result to a neighbor, who must now try to reconstruct the circle from the given area (Shell Centre, 2010, p. 7).

Further Reading

Baroody, A., Feil, Y., & Johnson, A. (2007). An alternative reconceptualization of procedural and conceptual knowledge. *Journal for Research in Mathematics Education, 38*(2), 115–131.

Shell Centre. (2010). *Concept development lessons: How can I help students develop a deeper understanding of Mathematics?* https://www.map.mathshell.org/pd/modules/2_Concept_Lessons/pdf/2_Concept_Lessons_Guide.pdf

Promoting a Thinking Classroom

- Arrange the seating to foster communication.
- Tell your students that particular lessons or parts of investigations are specifically going to involve talking about what it is like to work on solving a problem or posing a question, or about identifying a good way to use compare charts and pictures in words, or whatever else will be the focus regarding reflection on and communicating one's mathematical thinking.
- Motivate your students to talk about what they have done: give them reasons for this, or find out their own goals and help them to see that this discussion will meet some of these goals.
- Ask your students questions that would help them to think about and evaluate what they have done – not about the answer to a math problem, but about the processes to get to answers or new questions or some other aspect of mathematical work.

- Show your students that you are interested in what they have to say. Make sure that you use what they say to guide further activity in and out of the classroom so that what they contribute makes a difference.
- Paraphrase or reflect on what your students say to check that you have understood. This is important to demonstrate that you have heard what they say. But make sure you write down what they say in their own words as much as possible so that you are not changing it into your words.
- Ask questions to clarify what your students are saying.
- Involve all students in the discussion.
- Show unconditional acceptance of your students' thoughts and feelings. Let them know that it is acceptable to be wrong or incorrect (more importantly that misconceptions become correct conceptions when interpreted in the ways that they are likely to be thinking).
- Give your students feedback. Let them know that they are learning to do something that is difficult but very important.

Here is a recipe for creating a *culture of thinking* through communication in your classroom: Pour the above ingredients into your teaching (Adapted from Ginsburg et al., 1998).

It sounds good to increase classroom talk and get learners interacting with both the mathematics and each other. But how do we do this? There is a long tradition of using communication, whether reading, writing, listening and/or speaking, to promote 'mathematical literacy'. Ideally, these forms of communication promote *thinking*, according to Herb Ginsburg.

We can get lulled into believing we are doing a good job, because everything is going smoothly. Unfortunately, there can be very little learning even in a classroom that looks good. Because we want to feel like we are doing our job, mathematics teachers like to show and explain. But showing and explaining everything can too easily short circuit actual learning. The most common structure of a mathematics classroom finds a teacher who shows and explains. Then the class looks together at sample exercises and illustrative extensions of the original demonstration. Next, students practice on their own before the class comes together once again for the opportunity to ask questions about what was not understood. This format runs the risk of 'training' unthinking students in ways to copy the teacher, which is most efficiently accomplished by shutting down thinking, rather than struggling to understand and later reaping the benefits of that struggle. A thinking classroom is all about avoiding the excellent appearance of a good classroom that is hiding mindless mimicry.

Training in the best strategies for obtaining answers to textbook problems is definitely efficient, if the goal is to perform well on a test within a short period of time. And of course we are all pressed for time, needing to cover more material than we know is possible. But students who spend most of their time on rote practice miss out on the challenges and confusions that pay off with deep understanding and stronger self-confidence in their own abilities. Sure, it is easy to assume that students don't think. After a few days in any classroom, a teacher can start to believe that students do not, or cannot, think on their own. Yet, many students have learned over the years that shutting down their thinking is the best way to please adults. When they start to think, they can get into trouble. This is why we need to help them to think again in mathematics class. Canadian mathematics educator Peter Liljedahl recommends starting with hard puzzles and problems that push students to their limits. He encourages us to confront the fundamental passivity of classroom seating by trying new ways to arrange our furniture, and to use highly structured group activities that promote discussion, peer review, and iterative thinking. Dan Meyer shares his approach to "give less help". He redesigns packaged lessons to remove all the details of problems, so that students have to search for those they need, which is another way to say, *think* about them. The recipe in this list works.

Further Reading

Fuentes, S. Q. (2020). S3D: *Fostering and improving small-group, student-to-student discourse.* NCTM.

Liljedahl, P. (2020). *Building thinking classrooms in mathematics: 14 teaching practices for enhancing learning.* Corwin.

Meyer, D. (n.d.). *Less Helpful dy/dan blog.* https://blog.mrmeyer.com/

Five Dimensions of "TRU"-ly Powerful Classrooms

- The Mathematics Content
- Cognitive Demand
- Equitable Access to Content
- Agency, Ownership & Identity
- Formative Assessment

This PlayBook list is designed to help teachers, math coaches and researchers have conversations with each other (Baldinger et al., 2016). The idea is to bring people together so that they can share stories and come up with new ideas, instead of telling folks what research has determined to be "best practices" no matter what the situation. The goal is to experiment with your teaching so that, over time, more and more, you can be confident that you are *T*eaching for *R*obust *U*nderstanding of *Math*ematics. Talking with others about the ideas in this list, sharing samples of student work from different lessons, can help not only to make your teaching more creative, but also helps you to understand and appreciate how complex your work really is.

The Mathematics Content

Students often experience mathematics in school as a never-ending series of isolated facts, procedures and concepts that need to be rehearsed, memorized, and applied. The "TruMath" goal is for students to experience the mathematics as a discipline rich in meaning. This requires a teacher to plan ahead of time by most importantly identifying the key mathematical ideas behind whatever facts and procedures are the focus of a particular lesson or unit. Different teachers might identify different ideas – this is perfectly fine. It is simply important to have a sense of what *you* have artfully chosen as your big ideas. Once you have figured out the big ideas, you can plan ways to highlight connections among the skills and concepts, strategies for relating concepts to each other (not only within a single lesson, but especially also across lessons and units), and creative ways to engage students with the ideas in an active way. The general orientation of "robust" mathematics teaching is to facilitate the learners in their own making sense of concepts and ideas for themselves. By explicitly modelling authentic performance of important disciplinary practices (e.g., reasoning abstractly and quantitatively, constructing mathematical arguments and critiquing the reasoning of others), and then facilitating learners in their own independent engagement with these practices, you can then use whatever formative assessment strategies you have planned to check on how well this is working, whether you need to improvise some back-up plans to nurture this more, and so on. Useful questions to ask yourself include:

– How do mathematical ideas from this unit/course develop in this lesson/lesson sequence?
– How can we create more meaningful connections?

Cognitive Demand

Such an approach to mathematics teaching and learning emphasizes students as engaging authentically with important mathematical ideas, that is, not simply receiving knowledge from a well-prepared teacher. Partly, learning like this expects the students will necessarily need to experience "productive struggle", grappling with difficult concepts and challenging problems (See "Jo Boaler's Five Elements of Effective Struggle" in Chapter 3). As teachers, we often wish we could make things easier for students. Ironically, learning is easier when it often feels hard! Our goal is to help students understand the challenges they confront and persist in solving them, while leaving them room to make their own sense of those challenges. The speaker Dan Meyer talks about "being less helpful". This does not mean to step back and not do our job. It means that our job shifts toward helping students to reflect on their experiences of struggle in order to identify what helps them to move beyond getting stuck or not knowing what to do next, toward realizing that they can be the source of original ideas and inspiration for themselves and others. Useful questions include:

- What opportunities do students have to make their own sense of mathematical ideas? To work through authentic challenges?
- How can we create more opportunities for struggle?

Equitable Access to Content

Naturally we want each and every student to have the opportunity to experience rich mathematics, to go through productive struggle, and to emerge as mathematically delighted with their life in school. That's the hope. What is the reality? We work on some students we feel were left out last week only to note how others are not engaged this week. By checking with ourselves during planning and assessment, we can consider whether we have a good range of options for how a student can participate in the life of the classroom. Which students are participating in which ways? Who is consistently on the margins, and why? How are the classroom norms, the language expectations, the official and unofficial rules for how to behave in the room, influencing who can and does not participate? Useful questions include:

- Who does and does not participate in the mathematical work of the class, and how?
- How can we create more opportunities for each student to participate meaningfully?

Agency, Ownership & Identity

Agency, Ownership and Identity together have to do with creating a mathematics community in which students generate and explain their own ideas, respond to each other's ideas, ask clarifying questions of each other in order to better understand. When we reflect on a lesson or unit, it is helpful to include time to focus on how we know which students have engaged with the rich mathematics, and in what ways. We hope to document for ourselves with evidence that we have we seen students explain their own and respond to each other's ideas. When having conversations with other teachers, we want to be

able to share and compare what that looked and sounded like in specific cases. Useful questions include:
- What opportunities do students have to see themselves and each other as powerful mathematical thinkers?
- How can we create more of these opportunities?

Formative Assessment

The TRUMath approach asks a teacher to make sure that the instruction is responding to what the actual learners are thinking, and not merely to what we hope or assume is on their minds. Instead of presuming that a concept or skill has been covered or mastered, and instead of checking for correct answers to exercises, we can combine our assessment with the style of instruction that creates meaningful engagement by using instructional time to simultaneously be (formative) assessment time – to make our efforts to understand what students comprehend or not, to gain insights into their ways of making sense of the mathematics, while we also help them to be better communicators with and about mathematics with each other. Critical questions include:
- What do we know about each student's current mathematical thinking?
- How can we build on it?

Further Reading

Mathematics Assessment Project. (n.d.). *Teaching for Robust Understanding* (TRU). https://www.map.mathshell.org/trumath.php

Peter's Practical Critical Thinking Strategies

- Treat Mathematical Actors as Mathematical Critics
- Make a Choice, Pursue it, and Consider the Consequences
- Obsess about Functional Relationships
- Problematize "the Answer"
- Problematize the Pedagogy
- Understand Mathematics as Rhetoric
- Recognize Ourselves as Objects of Mathematics
- Perform Celebratory Archaeology

There was a time when mathematics teachers tried to 'teach' critical thinking. At other times, teachers might have thought that learning mathematics developed critical thinking (Adapted from Appelbaum, 2000). I have spent the last

30 years trying *not* to think this way, but instead, to constantly remind myself that critical thinking is a human thing. My students will go through critical thinking as a process as long as my organization of classroom activity does not stifle it, reward it, or distort it. In addition to the Recipe for a Thinking Classroom ("See Promoting a Thinking Classroom"), here are some out-of-the-box ways to creatively continue challenging your students.

Mathematical Critics

I used to have my students invent their own procedures and algorithms and to always search out another way to do a problem. The result was presentation of multiple perspectives on a single situation. This helped to establish mathematics as a humanly constructed technology of meaning which I hoped would lead to two results: (a) a view of oneself as a maker of meaning; and (b) a view of mathematics as made by people, and thus subject to the same critique as other human endeavors, according to criteria of value. Now I recognize that it is not enough to provide a forum of presentation or solipsism. Dewey admonished that a democracy access but the opportunity to be heard. Students in my class not only explain their strategy or procedures. They now have to use another person's strategy or procedure in a similar problem/situation and participate in a discussion that notes the strengths and weaknesses of each. Students are asked to identify a situation in which they would use each strategy offered (for example, to explain to a younger child, to impress a town council member during a presentation on a local issue, to calculate to be most sure of their result ... Now my students perceive mathematical thinking through multiple perspectives and can articulate a plausible reason for selecting each perspective over others in particular contexts of use.

Consequences of Choices

I used to structure my activities to include a collection of critical thinking skills, in order to facilitate my students' development and refinement of these skills: comparing, contrasting, conjecturing, inducing, generalizing, specializing, classifying, categorizing, deducing, visualizing, sequencing, ordering, predicting, validating, proving, relating, analyzing, evaluating, and patterning. No longer viewing my job as skill training, I instead recognize that my students come to be with varying inclinations to use skills of critical thinking when they are in school. I can take advantage of this human trait, and making critical thinking the context through which mathematics is learned. I provided open-ended situations in which students use those skills to draw a mathematical conclusion or accomplish a mathematical task. Iask students to design the questions and investigations themselves. They then need to *choose* one question or investigation to try out in an initial experimental probe, and report to the classroom community about what they did, found, wondered about, or what they got stuck on Here they must choose as well. We discuss whether their choices of question, investigation, timing and format of reporting were foo ones, and how they made these decisions.

Obsess about Functions

I used to collect data in explorations and support students' identification of patterns in the data, encouraging them to search for more than one pattern or to articulate more than one rule or description of the same pattern. Now my students are pressed to go further, by exploring the ways that changes in one or more categories of data are related to changes in others. I see this as establishing a cultural disposition for noticing, finding, looking for, or inventing functional relationships, because functions and other almost-functions are valuable mathematical objects of study in general – relationships between things are the essence of concepts.

Problematize "the Answer"

I was once satisfied with successfully facilitating students in sharing the different ways they obtained an answer, and in sharing their strategies for arriving at an answer as part of explaining why others should accept their answer. Later, they would re-describe how they arrived at their answer in stories without using any numbers or shape names. They would then be required to explain why their answer was important in terms of the original question asked. I now see that as merely a first step towards students offering several possible answers to the original question once they have completed calculations. They explain their reasoning about how their calculation could lead to different decisions about what to do, rather than simply think that a solution is "the answer". They might create a screenplay script that dramatizes choosing among three or more actions based on the calculations, helping the script's audience to appreciate the complexity involved in confronting calculations with conflicting criteria. Problematizing the answer can also be accomplished by making the answer into a question: I sometimes start with the answer and ask what might lead to it.

Problematize the Pedagogy

Some teachers offer options for how students will engage with material. They might use centers with different approaches or organize jigsaw cooperative learning around key components of a unit. Those are good ideas. However, I have found more engagement in response to a presentation of a particular topic or skill as a controversy among educators. I have students experience at least three different ways that people have thought of for beginning to learn about something. I then have students evaluate each approach (a) for themselves, (b) according to what they perceived as the logic of mathematics; and (c) how it helped them to connect to things in their lives. Pedagogies differ, and they influence how learners come to understand the material. Students can learn the mathematics as "examples" of this, rather than as the focus, and even meet the objectives more successfully. Students can then create their own books, puppet shows for younger students, rap songs for older students, presentations for panels of experts, and so on, based on their judgement about how to best communicate with the audiences.

Mathematics as Rhetoric

There was a time when I would be self-congratulatory about collecting examples of mathematics in the news, on social media and streaming on TV. I could use them to prompt investigations, or to illustrate concepts. I imagined this promoted appreciation of mathematics in everyday life, and maybe even provoked students' own alertness for math in the world, which they would pursue on their own, share with each other, and incorporate into their perspective on life. I now ask my student how the mathematics is represented and to consider why the author(s) chose that way to communicate over other possibilities. We might consider different scales on graphs, choices among graph and chart types, units of measure, even the use or not of mathematics in an argument.

Recognize Ourselves as Objects of Mathematics

I want to go beyond all of the great ideas for students as apprentices, creating their own mathematics, conceiving of themselves as creators of mathematics, using mathematics as citizens to make informed decisions, to contribute to their world through mathematics. Those are all wonderful. Yet there is a hubris to it. I now see that creators of knowledge need to understand the responsibility that goes with that. How does mathematics turn our world into a "mathematized world", and are there any cautions that go along with this? When we collect data on one thing rather than another, or only collect data on what is "data-ready" what about those intangible things, or other data that we have been ignoring? When an algorithm is used to make decisions, are we sure that the algorithm is good, just, fair, or responsible? Might different algorithms lead to different results? What would matter? People are studied with mathematics; how are we the subjects of mathematics, and how does it matter?

Celebratory Archaeology

I once used multiple forms of assessment. Performance tasks, informal observations, interviews, tests, portfolios, rubrics, self-assessment journals… they are all great. But I now find it crucial for students to see for themselves that they have learned particular skills and concepts, and also, more importantly, that they can start to see how they can use these skills and concepts in contexts that they have *not yet* experienced. I call the work archaeology because I want to cultivate the sense that these skills and concepts are treasures that we can uncover as we look back on recent work, that we need to sift through open-ended investigations and ephemeral conversations to realize that we really did collect some very specific things that we can carry into our futures. I devote substantial class time to looking back over recent lessons and units. We make lists of what students find important, facts they can organize in more than two ways, conjectures or arguments that believe key to a summary of the material. Finally, we collect a "good list" of investigations they wish they could pursue if they had more time (and perhaps they will, as part of optional future work), conjectures they think they could "prove" to someone outside of class, and questions for further discussion. An important part of this discussion is our consideration of how to decide whether something should be kept on this final list, and the need to keep the list to a reasonable length in order for it not to lose focus.

Another piece of this last item is that students start to see the skills and concepts as potentially transferrable to new contexts. Based on that final list, I bring new challenge probes and investigation provocations to the class for a short exploratory experiment in using those things we learned in entirely new ways. This is "practice" in looking for ways to use what we have recently learned in new ways. Transferability of learning is notoriously questionable. But maybe practice in trying to do it will help students improve their ability to try.

Further Reading

Appelbaum, P. (2008). Critical thinkers thinking critically. In *Embracing mathematics: On becoming a teacher and changing with mathematics.* Taylor & Francis.

Five Days of Better Question-Asking

- Avoid information collection by probing for students' explaining, elaborating or clarifying
- Allow plenty of time for thinking to nurture reflecting and justifying
- Facilitate students in making the mathematics visible
- Use focus instead of steering toward a target
- Encourage students asking each other

This PlayBook list is perfect for rotating your attention throughout the week: Start on a Monday making sure to *avoid information collection,* and use each day to practice until you use Friday for *encouraging students to ask each other questions*. Of course, the goal is to become more effective with your questions every day, but at first, one at a time is a reasonable approach.

Avoid Information Collection by Probing for Explaining, Elaborating or Clarifying

Because almost all of us naturally over-use questions that ask for simplistic information, *shifting away from short answers toward probes* that open up students' contributions into sentences and paragraphs can make a huge difference in our classroom discourse. Instead of "what is the formula for …" or "what does the exponent tell you?" it can be more rewarding for both informal assessment and for helping students understand important conceptual connections to ask a question like, "… as you added those arrows on your number line, what decisions did you make …" or "I know you didn't use a table to think about that, but could you show us how you would use a table to think about the same connections?" Modeling questions that support a clarifying story about thinking will help students think about the questions they can ask themselves as they work independently (See the "Georg Pólya's *How to Solve It*" playbook). Such questions take the form of, "What would you do if …" "What would happen if …" "Describe why you did …" "Compare … How are they similar? How are they different?".

Allow Plenty of Time for Thinking to Nurture Reflecting and Justifying

We often anticipate an answer, and it feels awkward to wait. But students don't feel this awkwardness, because they are busy thinking. They aren't waiting. Ten or more seconds is still not enough time. Given the need for deep rather than quick thinking (See "Jo Boaler's Five Elements of Effective Struggle" in Chapter 2), we should *slow things down*. Even if some students believe they have an answer to our question, it is worth encouraging them to think more carefully and deeply, to extend their thinking, and to probe for more ideas while others are just formulating their thoughts. We can *encourage reflecting and justifying* by waiting as students anticipate the follow-up question that has a form like, "How might you convince a skeptic?" (See "John Mason and Colleague's *Thinking Mathematically*" playbook), "How are you sure of that?" "Can you create and solve a problem similar to this one?"

Facilitate Students in Making the Mathematics Visible

Making the mathematics visible is less about 'showing your work' and more about discussing connections among mathematical ideas, describing mathematical relationships – whether in words, symbols or pictures, and placing the specific task at hand in a broader context. Interpreting an array of rows and columns in terms of how it could be thought about with multiplication and division, for example, extends that array into initially unconnected territory. Applying previously studied ideas about functions to a piece of a graph that is not a function when taken as a whole entity would be another foray into interesting connections.

Use Focus Instead of Steering toward a Target

Steering toward a target happens when we are anxious to get to our learning objectives. We are so committed to that objective that we miss the learning opportunities that pop up along the way. A certain answer or a specific strategy might be our main target. The way to sense that this is happening is to notice that we have slid back into information collection, followed by a higher order question at the end, once the target is reached. *Focus* results when we stop the information collecting and integrate reflection and justification questions into our discussions. In contrast with the signs of steering toward a target, we can be more assured that we are nurturing a focus if we hear students noticing things, when they are clarifying their thoughts for each other, and when our questions can be approached in more than one way.

Encourage Students Asking Each Other

The most powerful mathematics conversations are those where we feel like we are not actually teaching! Focusing questions are more easily generated when students are committed to understanding each other's ideas. Most classes need initial guidance in how to do this. The best way to make it happen is to model the challenging questions like, "Why do you think that?" or "Where did you come up with that?" or "Could you have used another approach to solving this?" By *encouraging students to ask each other the questions*, we keep engagement high and participation at a level that facilitates the connections so vital to significant conceptual understanding.

Jenni Way, a professor and researcher at the University of Sydney in Australia, organizes questions into four types: Starters, Mathematical-Thinking-Stimulators, Assessment-Questions, and Final-Discussion-Questions. Starters are open-ended yet provide a focus that stimulates a direction for their thinking by establishing a starting point:

- How could you sort these ...?
- What happens when we ...?
- Thinking Stimulators provoke and apprentice students in mathematical strategies of working:
- Can you group these using something you notice?

- What patterns can we see here?
- Is there a way to record and organize what we are noticing?

Assessment questions push students to articulate their thinking for others, enable them to describe (for themselves as much as for others) how they have been working after a period of time, and give them a chance to reflect on their work so far:
- What have you discovered?
- What made you decide to do it like that?
- How did you ...?

And final discussion questions pull individuals into a whole group so that they can compare, contrast, find commonalities that point to important reflection, and help students realize that they have done significant thinking:
- Who has a similar/different answer/method? How is it similar/different?
- Have we found all the possibilities?
- Do you think we have the best solution?
- What would you explain about this to someone who is not here today?

The National Council of Teachers of Mathematics summarizes some current recommendations. They include:
- Observing another teacher – focus on their questions, and to discuss together the impact of questions on the types of discourse taking place in each other's classrooms.
- Reflecting on your own questions – by recording your class and listening later, focusing on the types of interactions that follow your questions.
- Setting up classroom norms – to encourage students responding to your questions with open-ended stories about their ideas, but also to nurture students posing questions that challenge each other.
- Using a variety of assessment questions – incorporating think-pair-share, calling on students and encouraging participation by making sure that each student's contribution is used in a constructive way for everyone to learn from (See students as intellectual leaders in "Gloria Ladson-Billings' Six Habits of Highly Effective Teachers").
- Leaving a question unanswered at the end of a lesson or unit – which helps students over time to appreciate that mathematics questions aren't always answered quickly and easily, and sometimes are not solved or understood fully for years and years.

Further Reading

NCTM. (n.d.). *Asking questions and promoting discourse.* https://www.nctm.org/Conferences-and-Professional-Development/Tips-for-Teachers/Asking-Questions-and-Promoting-Discourse/

Reilly, G., Dietiker, L., & Anderson, N. (2022). *Good questions for math teaching – high school: Why ask them and what to ask.* Heinemann.

Schuster, L., & Anderson, N. (2022). *Good questions for math teaching, grades 5–8: Why ask them and what to ask.* Heinemann.

Sullivan, P., Lilburn, P., & Anderson, N. (2022). *Good questions for math teaching, K-5: Why ask them and what to ask.* Heinemann.

Way, J. (2014). *Using questioning to stimulate mathematical thinking.* NRICH. University of Cambridge Faculty of Mathematics. https://nrich.maths.org/2473

Worksheet Options

- Use 'Pretty Good' Worksheets
- Do Something Interesting with the Worksheets
- Beware of 'Fun' & 'Cute'
- Expand the Audience for the Work

Here is the challenge: each time you are about to use a worksheet, ask yourself, "Is this going to contribute to student learning, or is it mostly keeping them busy with practice?" (See David Kirshner's Three Types of Activity").

Use 'Pretty Good' Worksheets

Some worksheets are *'pretty good'* – those that facilitate students in the creation of graphic organizers, note-taking, data collection, or planning, scaffold activity that fosters thinking; those that provide practice in a skill or application that a student has identified for themself as a goal can increase confidence and lay the foundation for more creative work later. However, word searches, word scrambles, unrelated coloring and pasting, recall on vocabulary, labeling things, and so on are mostly designed to keep students busy in collusion with you to look like they are well behaved students.

Do Something Interesting with the Worksheets

What if you are in a school where the worksheets are assigned to you? Or, what if you feel obligated to have students complete the worksheets that are included in your curriculum materials, to send home worksheets so that families can see them or other otherwise compelled to use not-so-good worksheets? Then the next idea is to *do something interesting with them*. Some possibilities are: instead of just completing them, have students find more than one way to solve each problem, not yet practiced in class; describe a situation in which the exercises could come up in real life, and also describe the kind of decision someone would make after completing the math task, for each exercise; take the sheet home for homework and write a letter to another student about what they think is the most important idea to remember once finishing the worksheet. Even more interesting is to critique the worksheet: write a letter to the worksheet creator agreeing or disagreeing with their choices of exercise, and re-writing any uninteresting problems or tasks to make them more interesting, explaining how this makes them more interesting (or less challenging if it is felt to be too challenging); change the problems into more relevant problems with more realistic numbers or shapes, and work on those; use the exercises as first problems in a problem-posing activity by listing attributes, changing one or more, and pursuing an investigation prompted by this (See "Stephen I. Brown & Marion Walter's Art of Problem Posing"), or create a podcast where the exercises on the worksheet are used one by one, not to explain solutions, but to create the plot of a story. Have students organize and classify the tasks on the worksheet by attributes (See "Mason & Colleagues' *Thinking Mathematically*" and "Concept Based Lesson Genres").

Beware of 'Fun' & 'Cute'

Beware of fun or cute worksheets, or not-so-good worksheets in disguise. Coloring a pretty picture, completing a puzzle, or trading problems is hardly more meaningful than a standard worksheet. Putting each task on separate cards, passing them out, and having students move around the room explaining their problem to another student before trading cards and moving on to others can be fun, but is hardly better than sitting still and filling out the blank spaces. Splitting up the class into teams who compete to provide correct answers is a fun alternative to quiet drill and practice, but is basically the same thing but with speed attached (See 'deep thinking' in "Jo Boaler's Five Elements of Effective Struggle" in Chapter 2, and "Siane Ngai & the Aesthetics of Students' Contributions" in Chapter 6).

Expand the Audience for the Work

Students can *expand the audience for the worksheet* by translating it and their solutions into another language, sharing the translated sheet and explanations with a family member or neighbor, and sharing the experience with the class the next day (See "Celebrating the Brilliance of Multilingual Learners" in

Chapter 7). Perhaps you can have the students take a worksheet and re-write it for their classmates to use as homework, including tasks that they believe are easy, challenging, and more challenging. They can create a practice sheet for older students who once learned this material, so that they can refresh themselves through a review worksheet; here they need to think about what is most important, what the big ideas are, what the significant skills and procedures are, and so on; they should provide a teachers' guide with this information. They can compose a journal entry reflecting on the value of a worksheet, its main ideas, and the important procedures that it helped them to learn.

When students create their own worksheets, they show what they believe 'school is about' – they rely on the examples that have been most common in their years in school. They typically create mindless not-so-good sheets and are less likely to create 'pretty good' worksheets – what some might call *powersheets*. Discussing a worksheet as critics can help students to think about what forms of "practice" really do help them learn as opposed to filling time and keeping busy. As they become more refined in their analysis of what helps them to learn, they become more independent and autonomous mathematics learners, better able to create powersheets on their own, and more aware of what is important in mathematics learning.

Further Reading

Choose any other section of this book, find someone to share it with, and talk about it with them! Invite them for a cup of coffee, an ice cream, or a healthy stroll through a nearby park.

Storytelling & Literature Lists

Daniels-Type Literature Circle Roles

- Literary Luminary
- Connector
- Illustrator
- Soundtracker
- Question Collector

Decades of encouragement to use literature grounded in mathematics concepts and skills as an alternative to traditional lessons has left us continuing to ask, "But how do we do this?" One way is to guide learners in reading literature *as mathematicians* in small reading groups. The idea is to replace traditional lessons with reading, and then to observe that the experience has led to at least as many mathematics learning outcomes. Ordinarily, groups are better in theory than in practice. Teachers get frustrated as they watch one or two group members take charge, as other students wait for them to do the work; the watchers are frustrated because the take-charge members of their group haven't given them a chance, and the students taking leadership are frustrated that they always have to be the ones to take the work seriously and get it done. Differences in style create an inefficient and ineffective group. Training students in specific roles they can play in their group can help. However, commonly used roles don't seem to make enough of a difference, and the groups continue to be a lesson in what to avoid. Perhaps ongoing processing of the experience with the students could lead to more productive group skills. More powerful, it turns out, is to learn from productive group work in industry: each member brings a set of specialized skills and perspectives. Harvey Daniels and his colleagues have, over the years, found it effective to have students come together with such specialized ways of reading. In my own teaching, I find it highly useful to adapt the roles to be specific to behaving as a mathematician:

Literary Luminary

You identify key passages and quotes from the text that you want the group to analyze, understand, critique, and otherwise explore. You should be prepared to explain why you have made these particular selections, although you may not need to share this information during the discussion. Help people remember some interesting, powerful, funny, puzzling, or important sections of the book! You should prepare to read at least one thing aloud, dramatically with meaning; you might ask others to read other selections. During the discussion, ask others to find quotes or passages that support their interpretations or conjectures. Keep the group on task in discussing the book by bringing them back to the details of the text!

Connector

You make connections to other things we have read and/or discussed, as well as to specific personal experiences in your own life, and to anything else "outside of the book or our class". As you get to know your group-mates, you may be able to make connections to things you know about their interests and concerns as well. Also, look for connections within the novel: how are events, places, people, symbols, etc., related to each other? Any relationship between such things that you can find is a key to an important concept. Come prepared with your connections. During the discussion, get the others to make connections by asking questions (e.g., have *you* ever felt the way "…" did when "…"

happened in the book? Have you ever been in such a situation and found it turned out differently?).

Illustrator

Your job is "representation and analogy". You draw pictures, make charts, organize ideas in webs, and otherwise represent ideas in the novel to the group. Stick figures and flow charts are just as good as sketches or cartoons. Show your work without comment to the others in the group. One at a time, have them speculate on what your picture/etc. means, and to connect it to their own ideas about the reading. After everyone has had a say, you get the last word: tell them what your picture means, where it came from, or what it represents to you. Also, pay close attention to metaphors that are used – both by the author, and by the members of your group as they talk, because metaphors are valuable ways of representing important ideas. Point these out to your group, and ask them to consider the implications.

Soundtracker

Music can be a powerful force in a movie. Watch your favorite movie again, and listen to the way music can convey emotion, build suspense, or add to the excitement of a movie. Pretend your book is being produced as a motion picture, and you have been hired to create the soundtrack for the film. As the Soundtrack Producer your job is to determine what songs and artists would be a part of the movie's soundtrack. In addition, you will design an album cover for the soundtrack. To help you in the development of the soundtrack, answer the following prompts: What genre of music would fit this movie? Is it suitable for multiple genres? Explain. – Bring a playlist with music samples!

Question Collector

You are on a serious mission to pose questions. Prepare a list of questions that the group might discuss, and, whenever the group feels like they have run out of steam, offer one of your questions for discussion. Don't worry too much about details – help the group to talk about the main ideas. Often, the easiest way to do this is to get folks to consider what they thought about while reading: What was going on in your mind as you read that section? What questions did you have when you finished that chapter? Did anything about that character surprise you? Don't work for answers: generate discussion. Indeed, there may be no answers to your questions. Your ultimate goal is to get the group to end up with more questions than they started with. "What questions do we have now?" is your favorite thing to take notes on. Support others' efforts to ask questions, because they are doing your job for you! Collect questions from the group to bring to the whole class.

Literature Circle Playbook

- *Alternate between Role-groups & Book-groups* . In the role groups, students support each other in preparing their role for their group. They identify what works for that role. In the book groups, students take on their roles, and take notes to bring back to their role groups. The key idea is that students use both group formats to learn more about what makes a good version of their role – which, in the end, comes back to what makes someone good at that way of working as a mathematician!
- *Rotate students through the roles*. This helps them experience all forms of being a mathematician, helps them appreciate what others are trying to do with their roles, and helps them support each other in their book groups.
- *Discuss what makes someone good in each role, in small groups and whole class conversations*. Make lists of specific ways of working that are effective.
- *Discuss what makes for an effective book group, in small groups and whole class conversations*. Make lists of things that work and things to avoid in a group.
- *Organize short reflective sharing of "Appreciations"* in which students recognize for the whole class specific actions and choices that other students have done during the group meetings that led to positive outcomes.

Further Reading

Clark, L., & Holwadel, J. (2007). Help! What is wrong with these literature circles and how can we fix them? *The Reading Teacher, 6*(1), 20–29.

Schlick, K. (n.d.). *Literature circles resource center online*. http://www.litcircles.org/index.html

Storytelling Techniques for Math Teachers

- Know *Your* Story about Math
- Show, Don't Tell
- Try Binary Opposites, & then Avoid Them
- Change the Standpoint
- Add Standpoints
- Exploit Resonance & Wonder
- Change the Intended Audience
- Make It Not About the Math
- Make Students into Story Consultants

Traditional storytelling is the process of weaving language into a concrete narrative in order to create rich, believable experiences. Storytelling conjures images of a group gathered at the feet of an elder, transfixed by a magical tale. For mathematics education, storytelling can inspire us to think of our work as involving stories told through more than just language by a gesturing teacher or student. The language might include pictures and representations of mathematical relationships; situations in which mathematics makes a difference; it might take place more generally and abstractly, *about* mathematics as a subject or a specific mathematical concept, over the course of a year or a lifetime. The value of thinking about mathematics teaching *as storytelling* is that we can tap into the power of stories. Stories preserve human history and cultural identities. Stories seem to distinguish humans from other animals, they are a core of being human.

The first way to tap into all of this story-energy is simply to recognize that your teaching indirectly communicates a story about what mathematics *is*. As Kieran Egan (1986) proposes in *Teaching as Storytelling*, the ways we organize the experiences of learning any subject "tell a story over time" about the discipline. He further encourages teachers to exploit this by planning what story they want to tell. To put it plainly, the "story about math" that is commonly "learned" – or at least, suggested in popular culture – is that it is a boring collection of time-wasting challenges that are hard to remember and that make a lot of people feel stupid, while a few extremely smart people can use it to show how smart they are. But another parallel "story about math" is that it is important for doing science and for inventing technologies that make possible our modern way of life. What experiences with mathematics or through mathematics lead to these summaries about math? Pictures of typical classroom flash through our minds; cell phones, the internet and scantrons at supermarkets; statistics, diet and health advice; weather predictions.

Know *Your* Story about Math

What story *about* mathematics do you want your students to 'learn'? Egan says this is up to you! The provided curriculum gives you the pieces of the story – the characters are math concepts and skills, the setting is the classroom and the situations of word problems, the plot is how you sequence experiences so that they build characters, bring them together in situations, and lead to surprise, expectation, anticipation, suspense, mystery, and so on. If you want your students to believe that mathematics will make things possible in their lives, you can structure the plot to reveal ways that mathematics makes things possible. If you want students to believe that mathematics is beautiful and aesthetically rich, another version of poetry and music and dance, then your students need to go through character development (conceptual understanding, and standard algorithms) through the plot (sequential action that brings the characters together in different situations) so that the beauty and wonders of artistic immersion come to life. Stories are such an essential component of human experience because they work at the level of metaphors. Great stories tie together character and plot, resulting in stories that act as metaphors for the human experience. A good enough story about mathematics works in this abstract way as well. A first step is simply to try out what you think the story is that you want your students to experience, without really trying any specific way to make it happen. Just knowing your story will probably lead you to talk about and introduce mathematical things so that the story is implicitly told without your actually knowing when and how you are telling it.

Show, Don't Tell

In fact, the best storytellers *never directly tell*. They show the story behind the surface story through the characters and plot coming together. If you want your students to experience mathematics as powerful, you don't tell them this. You show them. You choose examples and challenges because the mathematics makes things possible that are otherwise not knowable or impossible without the mathematics. Another teacher more interested in mathematics as interesting because it is like a logic game would choose activities because they highlight this feature of mathematics. Right now, curriculum designers have chosen for you, highlighting mathematics as a collection of procedures for obtaining answers to other peoples' questions. The textbooks and worksheets and problem sets never literally say that this is what mathematics is. They indirectly show that this is the case. You can adapt those materials to show anything else you want to show. You are still using the materials, and students are still practicing skills, but if you redesign the conversations and activities, a different story can be told.

Try Binary Opposites, & Then Avoid Them

Kieran Egan recommended dramatizing your backstory behind the surface story through design based on binary opposites. Different storytellers could tell the tale of Cinderella with the same characters and same basic plot, the way they tell it might emphasize good versus evil (good Cinderella triumphs over evil stepmother and sisters, because good wins over evil); gender relations between men and women (the stepmother and Cinderella, and actually all of the women must marry a man to be able to survive, is such a patriarchy ok with you? – we never literally say this, but how we tell the story raises the question); or class structure of rich and poor (Cinderella and the stepmother have to marry a prince or at least a rich man to be able to survive in this world – we don't literally say this, but our details and how we talk about the plot and the characters and their emotions implicitly communicates this). Egan suggests a storyteller should choose which one of the binary relationships they want to use, and then build their character development and plot around it. If mathematics makes things possible is the message, then we create experiences where you are either powerless and lost or in control, and the mathematics makes the difference.

Binary opposites are limiting, though. And if teachers always taught with binaries in mind, then our students would learn another hidden backstory about knowledge in general: they would 'learn' that thinking can only happen through binaries. Is everything always right or wrong, black or white? No. There are often *more than two sides to a story*. So it is probably better to try out dramatizations through more than two categories. This is why most of the lists in this book include more than two items: Alice Pitt's Three Assessment Categories; Hannah Arendt's Three kinds of Activity and Four Strategies for Unlearning; David Kirshner's Three Types of Activity. Even mathematics itself has come to appreciate that there are commonly more than two kinds of truth: fuzzy logic allows for multiple truth-values for solving problems with open and imprecise spectra of data and heuristics. So we can specifically plan our instructional approach with such multiple categories in mind, in order to tell more nuanced and thus more powerful stories about mathematics through our curriculum.

Change the Standpoint

Standpoint is a critical choice for the storyteller. Who is telling the story? From whose perspective do we hear the tale? Imagine Cinderella told by Cinderella herself, instead of a narrator who is not in the story. Or the same story told by one of the stepsisters. That is a very different story. Even more unusual would be the same plot told with details that matter to a cockroach in the house. Writers use these ideas all the time. We can, too. Who is using this mathematics, in what situation? Why? How does that matter?

Add Standpoints

We can start with one standpoint, and then look at the same part of our story from another, contrasting standpoint. The mathematician using the mathematics or the mathematician as an explorer, traveling through unknown territory. The story can be understood from the standpoint of the mathematical object itself: adding fractions told by one of the fractions being added; sine functions described by a tangent function who is arguing with the cosine function; conic sections manipulated by a two-dimensional hexagon. In "Peter's Practical Critical Thinking Strategies", students become the objects of study, and tell the tale of the mathematics acting upon them, used to make decisions about them, and so on.

Exploit Resonance & Wonder

Museum curators use the storytelling approaches of resonance and wonder. Resonance refers to the ways that already familiar artifacts can reach out beyond themselves to act as touchstones for making sense of the world in general. On the one hand, students will bring their own associations with mathematical concepts and tools from previous experiences to whatever they are doing now. Those previous associations will resonate with whatever they interact with in the moment. This is like saying that listeners bring to a story their own associations with the things and events in the story, so that they are hearing a story unique to themselves. Previous experiences and associations resonate with what is in the story, and the feelings evoked in the story. In a museum, a curator knows that visitors will think that an artifact on display reminds them of stuff in their lives, and then also that whatever they experience in the museum can reach out into the visitors' lives in ways that might be mediated by the way that it is displayed. We can also exploit this by making sure that previous mathematical concepts and procedures 'reappear' throughout a lesson period or our school year, so that they will indeed resonate in ways that we, the storytellers, are controlling to some degree.

The literary critic and cultural historian Stephen Greenblatt (2018) contrasted resonance with "wonder", which refers to the 'wow-factor' of an artifact, its ability to stop a visitor in their tracks through its uniqueness, it special arresting demand for attention. Wonder is evoked either by the artifact or by the way it is displayed. Special lighting? Hidden around a corner? So, while resonance is powerful, we might want to keep a mathematics concept hidden until a special moment when it is revealed. Or we can save a particular method or form of representation for when it will make a startling impact tied to a specific plot or character development in our story (which, of course, we are continuing to show rather than directly tell).

Change the Intended Audience, or Make It Not about Math

The last three items on this list connect storytelling with other lists in this book. Using models and representations to communicate with a particular audience, inspired by Josiah McElheny's "Three Ways to Use Models" (Chapter 6) is a kind of storytelling. Shifting the audience might suggest shifting the chosen model. It might also involve a form of translation among representations or mathematical concepts, inspired by Hannah Arendt's use of translation as a form of unlearning, in "Hannah Arendt's Four Strategies of Unlearning" (Chapter 4). In Mason, Stacey and Burton's "convince yourself, a friend, a skeptic", one of the steps in "Thinking Mathematically" (Problem Solving and Posing Playlists), shifting the audience might require different details in your story about why you know something. Maybe something needs to be emphasized, or perhaps too much detail about certain things might turn an engaging story into an uninteresting or tedious retelling. Finally, in "Susan Sontag's Alternatives to Representation" (Chapter 6), the mathematics is turned into the "decoration". This is only an instructional strategy, and not a refusal to do your job. Of course we want to the students to learn the mathematics. It just might be easier, or more seriously learned if it is not the primary focus. We can use the mathematics to make sense of something else. If so, it is critical to the plot or character development as a minor character, not as a central character; or it is a significant feature of the situation in which the characters find themselves.

Make Students into Story Consultants

Finally, if the students are story consultants, they are learning how to look at the ways that mathematics is presented to others, how a narrator structures a tale to include mathematical and non-mathematical characters that develop and take on special features through a plot that requires situations, and which highlights relationships to communicate a story behind the surface story. Students can share story drafts to experiment and see the reactions of their audience, and then the audience can provide feedback. As writing teachers often say, the learning is in the revision process. So it will be powerful for students to rewrite what they want to communicate with others, change the standpoint of their story teller from one version to another, within the same story, or across different stories they tell. They will become skilled in recognizing what sort of story is being told, or what type of story is asked of them by an evaluator. In this last sense, they will gain the wisdom to psych out which of "David Kirshner's Three types of Activity" (Chapter 1) is being asked of them on a standardized test, in the routine daily life of a classroom, or by another student. They will also be able to provide the same sorts of feedback and wisdom to their teacher, to help with reflection on how things are going in the class overall, through their reactions to and feedback on the teacher's stories.

Further Reading

Egan, K. (1986). *Teaching as storytelling*. University of Chicago Press.

Greenblatt, S. (2018). Resonance and wonder. In *A museum studies approach to heritage*. Routledge.

Three-Act Mathematics
- Engage & Perplex
- Seek Information & Solution
- Discuss Solution & Its Implications

Three-Act mathematics is a popular approach to mathematics teaching designed to make the mathematics accessible and interesting to all students in a classroom. It starts with an "Act One" in which a video or other demonstration of a real-word situation invites students to share what they notice, and then what they wonder about the situation. They generate and collect questions they could try to answer about the situation. The *Act One* situation is intentionally open-ended without any 'helping details' that might be found in a typical word-problem exercise, so that the *Act Two* part of the structure can take place: given what the students wonder, they seek information that can help them to use mathematics in finding a solution. In a well-known TEDTalk by the American mathematics teacher and popular inspirational speaker Dan Meyer, he shares a video of filling up a water tank from a hose attached to a tap outside a building. The "wonder" is how long it will take to fill the tank. Watching the video is one way to find out – but it goes on and on and on and on … is there a less tedious way to find out how much time is needed? If only students could have an idea of the size of the tank, the amount of water flowing into the tank per minute, or other details, they might be able to quickly calculate the time. In fact, Dan got the idea for this video "problem" from his textbook, which provided all of these details. By following his strategy for three-act mathematics, which includes "providing less help" through scaffolding the procedures, he took away those details in the video … students needed to seek out the measurements, think on their own for how to use them, and eventually to work out the solution.

Act Three in this approach brings the class together in discussing the solution and its implications. It is similar to the generalizing found in the "Mason and colleagues' *Thinking Mathematically*" playbook, and shares much with the "Stephen I. Brown and Marion Walter Problem Posing" playbook: What if the tank were a different shape or size? What can we say in general about how to use the measurements of any tank to determine the time? Also: What if the water were flowing at a different rate? What if we needed to turn off the water and leave in 20 minutes, how much of the tank would be filled? How high in the tank would the water be?

Three-Act mathematics has taken off in the 2000s, so there are numerous websites where teachers are sharing their lesson ideas and materials. They can be the start of further investigations and probes, by provoking families of problems (See "Stephen I. Brown and Marion Walter Problem Posing", and "David Kirshner's Three Types of Activity" in Chapter 1). Revealing the solution in Act Three can also serve for some students as the community-bonding experience of a puzzle, in which they might not have thought of a solution on their own, but appreciate and understand the solution when it is explained (See the discussion of puzzles in "David Kirshner's Three Types of Activity").

Engage & Perplex

The open-ended quality of Act One is important in that it can help students to learn over time how to differentiate between questions that are possible to pursue with mathematics versus questions that are really not amenable to mathematization. For example, in the water tank problem, whether or not we want to fill the tank with clear or soapy water is more of an issue about cleaning the tank rather than about something that can be measured, such as time or volume. Similarly, the color of the water, the material out of which the tank is made, the weight of the tank, and so on, are not really possible to use in a directly mathematical way.

Seek Information & Solution

In Act Two, students can use information readily at hand, or ask for additional information. They might research this on the internet or by phoning a classroom mentor (See the discussion of a Community Advisory Board in "What Every Mathematics Classroom Never Knew it Needed", Chapter 10). Sometimes, the teacher might simply provide information as students ask for it. The new information might lead students to modify their original questions, and this should be noted, documented as important, and discussed in the Act Three whole-group conversations.

Discuss Solution & Its Implications

Act Three opens up possibilities for comparing solutions, identifying what they share in common, options for comparing exact calculations with estimates that were made during Act One, the identification of key skills used or learned, and finally, the creation of a new set of questions provoked by the work so far. What assumptions were made in order to get started? If we question those assumptions, do we open new investigations?

Further Reading

Beattie, E. (n.d.). *How to engage your students with 3-Act math tasks: Classroom organization & environment, math, teaching strategies.* https://www.mrsbeattiesclassroom.com/2019/01/engage-students-with-3-act-math-tasks.html

Ehlert, D. (n.d.). *When math aphpens.* https://whenmathhappens.com/3-act-math/

Meyer, D. (n.d.). *Dan Meyer's three act math tasks.* https://docs.google.com/spreadsheets/d/1jXSt_CoDzyDFeJimZxnhgwOVsWkTQEsfqouLWNNC6Z4/edit#gid=0

Meyer, D. (n.d.). *Math class needs a makeover.* https://www.ted.com/talks/dan_meyer_math_class_needs_a_makeover?language=en

CHAPTER 8

More Than Human Mathematics Lists

- Neo-SAMBA School Math
- Seven Ways Math Can Save the World

Neo-SAMBA School Math

- Science-Arts-Mathematics-Bio-Sustainability-Action as an update to STEM & STEAM
- Action & Community-Building Focus
- Fractal Assessment
- Recursive Planning & Action
- Curricular Surface & Depth

Mathematics teachers are used to being lumped together with science, technology and engineering, as part of a STEM curriculum, or more broadly in STEAM

© PETER APPELBAUM, 2024 | DOI:10.1163/9789004688469_009

with arts added on. When this happens, we worry about being watered down to the language of science, or into the tool of decision-making for engineering and design. What happened to the mathematics? Some of us would rather be grouped with language arts, because we see mathematics as an evolving literacy with reading, writing, listening, speaking and depicting. The mathematics is either the content of our communication, or the modeling and guide for our decisions. At other times, we might think of social studies and public policy being even more appropriate, so that we can integrate mathematics with data collection and social justice curricula. In SAMBA schools, *science, arts, mathematics, bio-sustainability and action are intentionally interwoven and constantly challenging each other* to generate authentic communities characterized by joy, creativity, and the courage to act on their convictions (Appelbaum, 2018). SAMBA schools are inspired by the samba schools of Brazil, community-based, intergenerational year-round schools that prepare for larger community events.

Action & Community-Building Focus

Traditional Brazilian samba schools practice and perform Afro-Brazilian dance and drumming. They are located in a neighborhood and celebrate Afro-Brazilian heritage and culture, in direct contrast to the mainstream education system. The term *escola de samba* has been said to originate in an early group that met in a school yard. There is no direct or formal instruction. Community members meet throughout the year to rehearse for big events and parades. Voluntary participation enables neighborhood development, healthcare, and shared creation of thematic performance plans, costumes, musical composition, choreography, parade floats, and more. Observers describe great fluidity between watching and dancing, across generations. The challenge sambas schools present to us is their conceptions of apprenticeship and leadership, observer and participant, knowledge and action. Might we create a samba-school-like experience with our own communities? Might they use mathematics as the drumming and dancing, with ideas generated across generations and between those metaphorically dancing or drumming with mathematics and those metaphorically in the audience for the mathematical dancers and drummers?

SAMBA schools pursue alternative geometries of curriculum. While some integrated curricula move in a linear fashion through various disciplinary orientations and modes of inquiry – in this case, that would mean starting with a science topic or approach, then shifting into an aesthetic and arts orientation, followed by mathematics, and so on – neo-SAMBA schools *interrupt the activity when it seems like they are stacking one subject on top of another.* Such activities do not mix and weave the modes of inquiry and fix them into stereotypical blocks of subject matter, like most of us expect in a typical school mathematics curriculum. It is hard to visualize what this means until you are involved as the mathematics representative in a SAMBA community-building school program, or starting one yourself. The key is to keep your goals in mind, that the group will stop and re-orient themselves whenever it feels like a linear stacking of one discipline upon the previous, and ask, "How are we building community through taking action?"

Fractal Assessment

One way to avoid the discipline-orientation that we are so used to, is to simply recognize that we have little to no experience working as mathematics teachers in a non-disciplinary program. In these new kinds of programs, it is worth thinking in terms of *fractals*, where each level of observation upon the group activities provides a self-similar picture. However we look at the activity in space (locations and movement) and time (moments and movement forward or backward through time), it resembles any other observation. Another strategy is to routinely reflect on the work so far, and use the pause to identify where, when and how each of the components of S-A-M-B-A have been actualized. This is different from stopping because you see that you are falling back into one of the disciplinary approaches: it is instead identifying characteristics of science topics or processes, arts techniques or sensibilities, mathematics or ways of working as a mathematician, bio-sustainability questions or topics or investigations, or social action processes with community building foci. Whatever is identified can be *recursively 're-entered' into the ongoing planning and community building* to provoke continued 'samba' work.

Recursive Planning & Action

An interesting feature of SAMBA schools is that they avoid what would otherwise seem like an obvious distinction between content and method. This is inspired by artists who dissolve boundaries among material, craft, subject, intention and function, such as Brazilian Vik Muniz, U.S.-American Sara Felder and South African William Kentridge. Muniz is well-known for a collaboration with 'garbage pickers' who work collecting recyclable materials in a garbage dump outside of Rio de Janeiro. He created giant portraits of the workers out of the garbage in which they worked, and then photographed the giant portraits to create art for auction in London, England. People treated as garbage were literally recreated out of the garbage, and turned into beautiful and valuable works of art. The art might be the garbage portraits, the design of the portraits, the photographs sold at auction, the new lives re-created from the money raised, or even the initial idea for the project itself as conceptual art.

Finally, a documentary made about the project (Walker et al., 2010) was a work of art communicating the processes, subjects, questions, dilemmas, and more. Felder uses poetry, juggling, and comedic storytelling in which the story and the objects juggled are props for each other, together creating challenges for further juggling and storytelling. As a self-defined 'queer juggler' we might say there is nothing specifically queer about her technique with the props – they fly through the air doing seemingly impossible things, much as the props of any juggler, queer or not. Perhaps her uniqueness in making the juggling part of a solo performance piece that incorporates juggling unusual objects, for symbolic purposes, in the larger context of her stories and humor, is queer, in the sense of her intention to be non-normative. By defying norms, that is, by pursuing fractal trajectories in a normative social space of comedic theater, her representations blur distinctions between the surface and depth of what an observer would say she is doing. Best known for his prints, drawings, and animated films, Kentridge often works with drawings, erasures and modifications, and re-filming, so the traces and memories are powerfully present in their futures. His art, for him, is its own form of knowledge, not simply supplementing the real world, and not purely understood in the rational terms of traditional academic disciplines. Although one might describe a linear set of experiences in a chronological listing of what he does in his studio, he acts as if linear thinking is abandoned and the material processes of the eye, the hand, the charcoal and paper become themselves the guides of creativity. In this way Kentridge believes that drawing has the potential to educate about the most complex issues of our era.

Curricular Surface and Depth

We need to use these artists as inspiration because there are so few examples of mathematics teachers participating in such programs. Yet, if we care about bio-sustainability (See "Seven Ways Math Can Save the World"), SAMBA school math or something like it may be our only hope. Analogous to the porous boundaries between watching and dancing in a traditional samba school, our neo-SAMBA schools would incorporate a similar fluidity between learning and teaching, or doing and knowing mathematics. The irony of a great communicator and explainer of mathematics directly teaching students is that it inadvertently trains the learners in skills and dispositions of a passive receiver of knowledge rather than a participant in the development of an ongoing community. Social movements expecting educative programs to change the world are disappointing in this way. Similarly, if we hope for mathematics education to 'change the world' we need to leap out of the surface of well-designed curriculum using best practices, into the depths of community building possible when there is a joint project of using mathematics as the art that creates community.

Further Reading

Kentridge, W. (2014). *Six drawing lessons*. Cambridge University Press.

Muniz, V. (1996). Surface tension. *Parkett, 46*. http://vikmuniz.net/library/surface-tension

Walker, L., Harlcy, K., & Jardim, J. (2010). *Wasteland*. Almega Projects.

Seven Ways Math Can Save the World[1]

- Designing Better Weather Forecasts & Climate Models
- Getting 'Bang for Buck' Out of Super Computers
- Making the Most of Renewable Energy Resources
- Preparing for Change
- Making Sense of 'Big Data'
- Developing New Technologies
- Making mathematics accessible to everyone

For a very long time, the typical mathematics curriculum has been organized around categories like algebra-geometry-statistics-calculus, or whole numbers-fractions-decimals, or algebra-analysis-topology-logic ... This tells a certain story behind the daily story of school mathematics (See "Storytelling Techniques for Math Teachers"), one of a static collection of things other people figured out that a student can learn. Maybe it's time for another organization? This is not a new idea: there have been reform efforts to reorganize the curriculum around themes of application, core principals and threads across subfields, and so on. It's not new, but it may be that we have the only resources that can save the planet if we would only organized the story behind any organization of our curriculum so that it implicitly communicates a different message: "Mathematics is important for saving the planet!" and the associated story, "If you learn mathematics, you can use it to save the planet!".

A way to do this is to have a continuous subtext in your choice of examples and problem-solving activities that demonstrate mathematics as contributing to one of the seven applications in this list. Just as algebra, geometry, probability and statistics, generalization and specialization, problem-solving and posing, and so on are present at all levels of school mathematics, preK-university, we can start now on our own to make these seven forms of mathematics present at all levels of schooling, in age-appropriate versions.

Weather Forecasts & Climate Models

Accurate *weather forecasts* predict when and where extreme weather may strike. *Climate projections* are key to identifying weather patterns changing on a longer time scale. Global warming means increased extreme weather events all over our planet, and life is increasingly affected. Weather and climate prediction uses a lot of mathematics and has made enormous strides. Weather forecasts use computer models to solve the complex equations that simulate atmospheric interactions. Jet streams and rain clouds can be modeled. Mathematicians play important roles in modeling with equations that describe the atmosphere, taking into account variables like temperature, pressure and humidity. Global Circulation Models (GCMs) describe the interactions between oceans and atmosphere to look at what the average conditions could be in decades to come. What data people use to create these models, how to communicate what the models are indicating, and so on, are all significant opportunities to study local and global weather in school.

Getting Bang for Buck Out of Super Computers

Computers modelling weather and climate as well as computer models for many other situations keep growing in power as new technology and programming evolves. However, power and processing speed are not the only ways that mathematics is involved in computer modelling. It also helps with the design of new forms of computing like quantum computing, and by inventing new algorithms used in computer programs. Innovations now allow a computer to automatically zoom its attention on its own without human direction, onto areas where the weather is particularly interesting (such as around storms) for example. Optimizing computers performance can also reduce their energy demand. Even modest efficiency gains could have a massive impact on overall energy consumption.

Making the Most of Renewable Energy Resources

Energy sources are significant contributors to climate change. Renewable energy sources are key to a future low-carbon world. Optimizing locations for wind and solar farms, designing layouts for tidal and wind turbine arrays, and even predicting energy use or need all use mathematics. Indeed, probability and uncertainty can be studied with mathematics and then communicated to the general public and policymakers so that they better understand heatwaves, floods, and other changes in weather patterns.

Preparing for Change

So far, this communication about *preparing for climate changes* has been left up to scientists and journalists. Mathematicians can help to make representations of the information comprehensible so that health and community can be saved beyond the length of our students' lives into the far future. And students can learn how to do this as well. Businesses also need detailed information to plan around the effects of climate change. The food industry is dependent on agriculture, which needs advanced warnings about drought or rainstorms. New systems of efficient transport can move food and consumer goods around the planet without destroying the environment. Predictions on where around the globe the risks are more likely can support better planning. Mathematica simulations can guide such planning, by testing out the long-term consequences of certain choices being made over and over again (outputs in a function become the inputs for the next iteration in such models). In this way, carious scenarios can be observed in the simulation. Mathematics is involved in phone apps that can help people know about the health quality of the air in one walking path versus another, the risk of infection on one form of public transportation on a given day versus another, and more. Personal decision-making will be relying on such apps, designed through algorithms that use big data in real time, more and more and more.

Making Sense of 'Big Data' & Developing New Technologies

Many of the examples require huge *data* sets for the models. Collecting billions of pieces of data in environments, from ice sheets to cities, can deliver precious insights into our planet's physical processes, human behavior and everything between. Climate scientists rebuild the history of our planet's atmospheric composition by analyzing the tiny bubbles trapped in ice records, in order to anticipate the scope of future changes. But without the statistical methods that mathematicians bring to analyze this data and assess its reliability, the information has less value. At the same time, *new technologies* are key to a low-carbon future. Carbon capture and storage, for instance, could safely lock away greenhouse gases emitted by fossil fuel-fired power stations, and is likely to play a key role in averting dangerous levels of global warming. Detailed mathematical models make this research possible by using sophisticated logistics methods, network analysis, statistical modelling and many other mathematical tools.

Making Mathematics Accessible to Everyone

Mathematics has a bit of a "decoration problem" in all of this (See "Susan Sontag's Alternatives to Representation" in Chapter 6). It all sounds like science, technology and engineering, and not so much about the embedded mathematics. As noted in other lists of this book, though, when mathematics is the decoration, students learn more mathematics more deeply, and are more likely to continue studying mathematics beyond when it is required of them. So maybe it's not such a problem. Here we are merely suggesting that you try googling these things to create your own versions of examples in your class.

It can be useful to engage students in local examples of such mathematics. The Green Schools Initiative (n.d.), for example, suggests facilitating conversations with your students early in the school year about changing habits and creating policies in their classroom and school. They maintain an updated collection of topics and resources. However, global corporations have kept people distracted in their local efforts like recycling while their own uses of energy and management of waste are the far more horrific contributors to our current climate crisis. So we need to move students beyond the local into the sphere of global activism related to these issues, and the use of mathematics, data, and projections based on simulations to effect political changes trans-nationally.

For a global set of 17 sustainability goals, the United Nations Department of Economic and Social Affairs currently lists 17 categories that can organize student projects (Table 3). Adopted by all United Nations Member States in 2015, these Sustainability Goals provide a shared blueprint for peace and prosperity for people and the planet, now and into the future. They recognize that ending poverty and other deprivations must go hand-in-hand with strategies that improve health and education, reduce inequality, and spur economic growth – all while tackling climate change and working to preserve our oceans and forests. Even these sustainability goals have been heavily criticized as minimizing the contributions of global corporations and the fantasies of never-ending economic production. Our attention to the more-than-human lists may support students in going beyond them.

TABLE 3 United Nations 17 sustainability goals

No Poverty	Renewable Energy	Climate Action
No Hunger	Good Jobs & Economic Growth	Life Below Water
Good Health	Innovation & Infrastructure	Life on Land
Quality Education	Reduced Inequalities	Peace & Justice
Gender Equality	Sustainable Cities & Communities	Partnerships for the Goals
Clean Water & Sanitation	Responsible Consumption	

Note

1 This list is based on a website by "Tom" and "Paula", who were PhD students at the Mathematics of Planet Earth Centre for Doctoral Training in 2016 (https://granthaminstitute.com/2016/05/31/seven-ways-maths-can-save-the-world/), when the Grantham Institute hosted an event, "Numbers to the Rescue: Can Maths Save the Planet?" (https://www.youtube.com/watch?v=cdx4Vm2iSyI).

Further Reading

Green Schools Initiative. (n.d.). *Tips for teachers: Start the school year off green!* http://www.greenschools.net/article.php-id=494.html

Pfaff, T. (n.d.). *Sustainability math: A quantitative literacy and mathematics resource.* https://sustainabilitymath.org/

Royal Meteorological Society. (2021). *Maths for planet earth: Climate based questions for students and teachers.* https://www.metlink.org/maths-for-planet-earth/

United Nations Department of Economic and Social Affairs. (2016). *The 17 goals.* https://sdgs.un.org/goals

CHAPTER 9

Excellence & Flourishing Lists

- Lists of Virtues
  - Aristotle's Virtues
  - Theoretical & Practical Virtues
  - Virtues Inspired by Judith Butler
  - Afrofuturist Virtues
- Shopping Lists
  - What Every Mathematics Classroom Never Knew It Needed
  - First Aid Supplies for the Mathematics Teacher's Body & Spirit
  - A Short List of Mathematical Games You Might Have Overlooked

Why study mathematics? There are more responses to this question than anyone can begin to say. There are practical reasons, academic reasons, personal growth reasons; there are beliefs that mathematics trains us in ways of thinking, that it connects us to spiritual and higher powers, that mathematics

© PETER APPELBAUM, 2024 | DOI:10.1163/9789004688469_010

is an astonishing body of human achievements deserving of respect; there are connections to other subjects that make mathematics indispensable. Mathematician Francis Su (2017) summarized all of these in his address upon retirement as President of the Mathematical Association of America. "Because how you answer", he said, "will strongly determine *who* you think should be doing mathematics, and *how* you will teach it".

"So if you asked me: why do mathematics?" continued Francis Su, "I would say: mathematics helps people flourish. *Mathematics is for human flourishing*".

What he explained is that a well-lived life is a life of human flourishing. And he urged us to think of mathematics teaching and learning as facilitating that flourishing. How do we use mathematics to generate human flourishing? Through excellence. Or, as the ancient Greek philosophers said when they were talking about excellence, 'virtue'. The philosopher Jennifer Frey (2001) writes that a flourishing society needed good citizens who not only display civic virtues, such as civility, friendliness, and justice, but personal virtues, like courage, wisdom, and self-control. Education, she explains, has cultivation of virtue at its core. Sometimes we confuse education with training, which transmit skills and information for the sake of work. But *education*, and that means *mathematics education*, makes learners into noble people, flourishing folks, who contribute to the flourishing of others. By extension, I will add other more-than-human members of the planet and the universe. We do not flourish apart and alone, but together in common associations.

We're not used to talking about virtues as mathematics teachers, but if we start doing just this, we will, it seems, *flourish while helping our students flourish*. We do this by "being excellent" – no wonder my favorite movie is *Bill and Ted's Excellent Adventure*! (Herek, 1989). The lists in this section also include self-care and essential supplies for the pursuit of excellence. Merely studying the virtues and collecting the essential supplies is not enough. However, study can help foster moral development. Our minds and lives tend to take the shape of what we think about throughout the day. Su mentions play, beauty, truth and justice as key reasons for the pursuit of flourishment, terms that have appeared throughout the rest of this book.

Further Reading

Su, F. (2017, 8 January). *Mathematics for human flourishing*. Web Blog Post. The Mathematical Yawp. https://core.ac.uk/download/pdf/70982118.pdf

Lists of Virtues

Aristotle's Virtues

- Temperance
- Liberality
- Magnificence
- Magnanimity
- Patience
- Truthfulness
- Wittiness
- Friendliness
- Shame
- Justice
- Modesty

The ancient Greek philosopher Aristotle defined virtues as characteristics and tendencies in an individual that causes them to act in a particular manner. We inculcate virtues by patience and by emulating 'moral exemplars' until it has been internalized. We become just by practicing justice, truthful by practicing truthfulness and honesty, courageous by practicing courage unless it becomes a habit, unless these virtues become an extension of ourselves.

He also believed each virtue is a "golden mean" between the vice of excess and its deficiency. When talking about courage, for example, if we possess the vice of deficiency, we will not have the courage to overcome any obstacles that cross our paths but if we have the vice of excess. We would then be courageous to the extent of foolishness. Aristotle regards both traits as virtuous. A virtuous person, according to him, would know exactly where and when to be courageous without bringing danger upon themselves or others.

Aristotle's 12 Virtues

Courage – also known as fortitude. Aristotle says courage is "observance of the mean with regard to things that excite confidence or fear, under the circumstances which we have specified, and chooses its course and sticks to its post because it is noble to do so, or because it is disgraceful not to do so". It is a middle ground between cowardice and foolhardiness. The courageous person is aware of the dangers that are to come his way but chooses to face them, nonetheless.

Temperance – This is one among the 4 cardinal virtues. By this, he means moderation or self-restraint to a decent extent. It lies between overindulgence and abstinence. For him, the man who drinks too much and the man who never drinks, are equally insensitive. Controlled drinking is what he would call virtuous.

Liberality – The goal of this virtue is to reach the golden mean between giving all you have and being a pinchpenny. This is the virtue of charity.

Magnificence – This is the virtue of living extravagantly. It is the midpoint between vulgarity (gluttony?) and severe austerity (depreivation?). Aristotle deems that man virtuous who does not abide by a code of severe self-discipline but knows where to draw the line.

Magnanimity/Pride – This is the virtue related to pride. The golden mean of this virtue is to recognize your abilities and have a sense of self-worth, but this self-satisfaction should not be raised to the level of extreme narcissism and delusions of grandeur. This virtue enables you to act according to your self-worth and strive for greatness.

Patience – Again, this virtue teaches and conditions you not to lose your temper very often. However, a patient man knows when to actually get angry. The whole art is to strike a balance between the two extremes.

Truthfulness – This is the virtue that advocates honesty. The man who is truthful lies between the vices of being an effortless liar and undiplomatic.

Wittiness – This virtue relates to a good sense of humor and intellect. The golden mean of wittiness is in between boorishness and buffoonery.

Friendliness – Friendship might be considered as a waste of time by many of us but you would be surprised to know that Aristotle assigned a very important position to friendship in a happy and virtuous life. The midpoint of the virtue of friendliness lies between being too friendly and being totally ignorant about friendship.

Shame – Again, the golden mean of this virtue is situated somewhere between being shameless and being too shy. It is all about understanding when and about what to be shameful. This is especially applicable when a person has committed a moral or social error.

Justice – Justice is another cardinal virtue. Aristotle says that- "It is in justice that the ordering of society is centered" and quite justifiably so. He believes that fairness and lawfulness should always prevail in every society and that being said, every human being must demonstrate equality in their actions and behavior and redress inequality and unjust practices. It lies between being selfish and being completely selfless.

How can we use Aristotle's virtues with our students? One possibility connects with "Susan Sontag's Alternatives to Representation" (Chapter 6) and with "Storytelling Techniques for Math Teachers" (Chapter 8), by using mathematics as the "decoration" for discussing virtues or by making the mathematics into a minor character or plot device of a story about something else. As informal assessment we should facilitate classroom reflection on recent mathematical work, and ask, either one at a time or all at once, how anyone has recognized one or more of the twelve virtues during mathematics lessons, homework, or other mathematical moments of life. Although this seems like something that is not part of the required curriculum, we know from the discussion of Sontag that mathematics applied in this way, as decoration, or from the discussion of storytelling, that mathematics used as a minor character or plot device can actually be a more efficient and meaningful way to facilitate learning of the mathematics itself.

Another way to use these virtues is to make them into a list similar to "Susan Ohanian's Ten Foundational Questions" (Chapter 3). They are not yet in question form, but simple tweaks and they all are:

- When and where have my students demonstrated this virtue with mathematics?
- How can I promote this virtue in my students, which of course is nothing other than *excellence*?
- How can I facilitate my students in recognizing any one of these virtues during mathematics work in themselves and others?
- How can I help all of us in the room to use mathematics in ways that promote any one of these virtues in ourselves and others, to use mathematics in ways that promote flourishing?
- How can I promote the use of mathematics in reducing forms of non-virtue, that is to minimize unethical acts?

Try it. The results will be astonishing – you are opening up a cabinet of "Resonance and Wonder" (See Storytelling Techniques for Math Teachers").

Further Reading

Marbán, J., Fernández-Gago, J. (2022). Mathematical problem solving through the lens of ethics and Aristotelian attitude: A case study. *Mathematics, 10*, 2565. https://doi.org/10.3390/math10152565

Theoretical & Practical Virtues

- Theoretical Virtues: Simplicity, Fruitfulness & Accessibility
- Practical Virtues: Permanence, Reliability, & Autonomy
- Mathematicians' Helpful Virtues: Flexibility, Adaptability & Caution
- Mathematicans' Background Virtues: Courage, Honesty & Restraint

"Virtues" sure sound *esoteric*. Crazy. What are they doing in a book like this? There's actually a whole world of people who have found practical and important discoveries through study of virtues in mathematics and mathematics education. There is even a special issue of the journal *Synthese* with contributors discussing theoretical and practical virtues! (Aberdein et al., 2021). For example, *Simplicity* is a powerful virtue in mathematics. The less complicated an explanation or proof the better. The fewest things involved the better. Elaborate and complicated explanations are frowned upon as lacking elegance. *Fruitfulness* is another virtue. If we know 'this' about a mathematical property or object, does it lead to results for *other* mathematical things? Does it help to solve problems in seemingly unrelated mathematics? Does it lead to a whole new set of interesting investigations? If so, then it is "fruitful". *Accessibility* to a certain audience is a virtue – does this mathematical thing or mathematical discussion make an impact on the audience? Is that impact because of its beauty, brevity, depth, elegance, explanatory power, fruitfulness, generalizability, purity, or visualizability?

Permanence, reliability, autonomy, and *consensus* are often suggested as 'practical virtues' in mathematics, because they support *progress* in mathematics as a field of study. Tamsin Meaney argues that mathematical thinking teachers us all to thrive by dwelling in uncertainty. Mathematicians themselves might enact virtues. Georg Pólya (See "Problem Solving & Posing Lists", Chapter 8) suggested three virtues that help a mathematician (labels proposed by me, not Pólya, but the spirit is there):

- *Flexibility* – a readiness to revise any one of our beliefs
- *Adaptability* – we change a belief when there is a reason to do this
- *Caution* – we don't jump to change willy-nilly without a reason

Pólya (1954) further identified special, background virtues of mathematicians that help them demonstrate those first three:

- *Courage* – Challenging your own intuition, your own assumptions, and especially those of others, demands serious intellectual and emotional courage!
- *Honesty* – Being honest with yourself about what you know and believe, and what you don't, is really tough. Sticking with an idea when there is evidence against it is dishonest, pretending you understand when you have a question or need something explained is dishonest.
- *Restraint* – This is related to Aristotle's "patience" but also emphasizes the wisdom to make sure about things. It would be foolish to go along with others without checking for yourself whether you agree with their results or their methods. Of course, the pressure is sometimes too much because we might get tired of thinking, need a break, or just do not have enough time to slow down and reconsider every little thing. Yet Pólya stressed the value of putting work aside and coming back to it the next day.

Virtues are a cool way to talk about the work in mathematics class. Instead of focusing on correct answers, sharing strategies for the sake of comparison, or memorizing useful algorithms, a class can either discuss the relative presence of a particular virtue in the (mathematical) activity they witness together, or they can use mathematics to act in the world in ways that exhibit one or more virtues. Here once again the mathematics might be the "decoration" for the virtues (See "Susan Sontag's Alternatives to Representation"). Or, discussion of virtues, theoretical and practical, might be an alternative to exercises, puzzles and probes (See "David Kirshner's Three Types of Activity"). Either approach uses mathematics to construct a community of mathematicians doing important thinking (See "Promoting a Thinking Classroom" and "Peter's Practical Critical Thinking Strategies").

Further Reading

Aberdein, A., Rittberg, C. J., & Tanswell, F. S. (2021). Virtue theory of mathematical practices: An introduction. *Synthese, 199*, 10167–10180. https://doi.10.1007/s11229-021-03240-2

Dubbs, C. (2020). Whose ethics? Toward clarifying *ethics* in mathematics education research, *Journal of Philosophy of Education, 54*(3), 521–540. https://doi.org/10.1111/1467-9752.12427

Pólya, G. (1954). *Mathematics and plausible reasoning*. Two Volumes. Princeton University Press.

Virtues Inspired by Judith Butler

- Recognition
- Response to Suffering
- Interpretation

The U.S.-American philosopher and professor of rhetoric, Judith Butler, has never, as far as I know, discussed, or offered an analysis of mathematics or mathematics education. Nevertheless, we can learn from their ambivalent perspective on ethics. Butler typically dissolves opposites into more nuanced alternatives that make the opposites less relevant than other aspects of human experience. They are famous for their early work on gender, in which they "troubled" most attempts to define gender expression and gender identity by proposing that gender is neither expressed nor an essential quality, but instead interpretations about us when we or others experience assemblages of our actions in the world. For example, in each moment, our clothing choices, hair styles and colors, voice, gait, and anything else together spark potential interpretations. In other moments, or in other places, the same or different combinations can provoke different interpretations. Their explanation for this is that gender is not a performed and fixed identity, but perpetually and constantly, fluidly changing or re-establishing itself as something they calls "performative". We can use this idea for anything that might feel stabile in our classroom. Is an individual student a 'good' student, or a student with a positive relationship with mathematics? Well, there are collections of specific behaviors, gestures, things they say and do not say, ways of interacting with other students, and so on, that together might be interpreted as the presentation of such an identity. For a fan of Judith Butler, such combinations are always changing, and do not represent a fixed essence of that individual. We can "trouble" these identities, making any moment the spark of transformation or reaffirmation of that particular combination and its interpretation. Similarly, a triangle is not always the performance of Euclidean two-dimensional line segments in a specific collection of attributes. Change any small thing about the space, like place the lines on a sphere or a cube, and we have different ways of understanding those individual pieces of the shape. Any mathematical idea or procedure is understandable as performative, rather than a performance of a Platonic, ideal thing. The list in this section extends Judith Butler's ideas to virtue ethics.

Recognition

Recognition for Butler is mostly about the ambivalence of people recognized as human. They often ask whether a person or a group are recognized as human in particular situations. Immigrants, migrants, members of the LGBTQIA+

community, and so on, are sometimes erased from conversation or policy, making recognition primarily a positive thing. One needs to be seen as present in order to experience the dignity that accompanies one's humanity. We can use this directly in our classrooms, and ask ourselves how we are making sure that each student is recognized by others as present, and indeed, important for the ongoing development of our classroom community. Do we make sure that their ideas and questions influence the discussions? Are they intellectual leaders in our classroom? (See "Gloria Ladson-Billings' Six Habits of Highly Effective Teachers", Chapter 7) Sometimes our classrooms are unwelcoming to particular students, because of disabilities or accommodations they need for full participation; this structures such students as outside the norm, as different and not a fully human member of the group. Can we re-organize our classroom and our chosen activities so that full participation of all is possible?

Butler's ambivalence is helpful in this sense, because they worry that recognition is often accompanied by forms of oppression or violence. This may seem extreme to think about in a mathematics class, but there is a way in which symbolic violence is common in school. When we "recognize" someone or some group as different in some ways, this can lead to blaming the victim actions, or singling them out for forms of special attention that are more harmful than good. Reflection on these things seems really important!

Response to Suffering

"Response to suffering" is a significant piece of Butler's theories, because it has to do with whether we are even aware of someone's difficulties, and what we do when we do become aware of them. First of all it requires recognition, since a lack of response is basically a lack of recognition. If we treated them as fully human, we would be obligated to alleviate suffering. What constitutes suffering in mathematics class, though? – finding something challenging is not suffering. Feeling unhappy or shameful about being incorrect is a kind of suffering. The response should be to facilitate over time a desire to use incorrectness as a sign of progress soon to come. Exclusion from the conversation is a form of suffering. Once we recognize this is taking place, we can guide the class toward forms of conversation that include everyone.

Interpretation

Interpretation is something Butler applies from Susan Sontag (See "Susan Sontag's Alternatives to Representation", Chapter 6). Sontag and Butler were originally thinking and writing about photography compared with paintings and other representations. Wartime photographs seem at first glance to be a very clear depiction of reality. In this sense, they leave interpretation up to the viewer. One step back from this sense and we realize that a photographer made choices about when, what, and from what angle to take the photograph. So the photographer used the photograph to make an interpretation. Butler notes that photojournalists specifically did this for much of history, but this doesn't happen anymore, because governments "embed" journalists in ways that control what can be communicated to the broader public. So the photograph is now a simple image of what the journalist was allowed to see. Sontag had written about paintings as always interpreting, and never having this sort of problem. Butler added a layer of interpretation: the person taking the photograph is part of what is taking place.

For a mathematics classroom, the mental and physical "pictures" of what is happening are like the photographs that Butler and Sontag are thinking about. We, the teachers, 'take these pictures' and in this way we are part of the action: that happened 'so that we could take the picture'. What did we decide to collect as student work samples, or what student comments did we jot down on our notebook as anecdotal informal assessment during small group work? If we collected that stuff because we were directed to do so by a school policy, teachers manual, or core curriculum expectation, then we are like the embedded journalists. To counteract the loss of interpretation from embeddedness, we need to establish routines of collecting data outside of the expectations, to bring an "independent lens" to our pictures of classroom life.

We can also bring these virtues to life for our students but encouraging the use of mathematics for enacting the virtues. Simply asking, "Okay, now that we found three ways to answer that question …" or, "Looking back over this last month's unit …" "how can we take what we learned and apply it to making sure that no one anywhere is denied the recognition and dignity of being a member of humanity?" or, "What picture of that situation was supported by our uses of mathematics, and what possible interpretations were avoided through the mathematics?" Connecting to problem-posing (See "Stephen I. Brown and Marion Walter's Art of Problem Posing", Chapter 8), we should encourage students to compare and consider the implications of problems being posed for them (erasing their individuality and participation in the process), and the nature of mathematics constructed by them by their choices of problems, methods, and criteria for interpretation.

Further Reading

Butler, J. (2007). Torture and the ethics of photography. *Environment and Planning D: Society and Space*, 25(6), 951–966.

Butler, J., & Taylor, S. (2010). *Examined life*. Video. https://www.youtube.com/watch?v=koHZaPkF6qE

Afrofuturist Virtues

- Reconstitution
- Refraction of Imagination, Technology & Liberation
- World-building Retrodiction
- Collaborative Appropriation with Community Elders

Tricia Rose, an American sociologist and professor of Africana Studies at Brown University in the United States, once suggested that Black women would change the patriarchal look of a cyborg. "If we had hordes and hordes of women who were paid to sit around and reimagine the science fiction genre, they might treat technology differently, placing it in a different relationship to the organism, and then what would cyborgs look like?" (Dery, 1994). Art curator and Afrofuturist Ingrid LaFleur seeks to imagine "possible futures through a black cultural lens" (Womack, 2018). Zamalin (2022), a professor of political science at The University of Detroit Mercy in the United States, is interested in how some artists and scholars undo in fiction what is socially familiar and naturalized in practice and in history. What they all have in common that is useful for thinking about mathematics teaching and learning, is an inversion of conventional thinking about oppression, stereotypes, powerlessness, or a fatalistic reckoning with harsh realities. In the spirit of Black Joy (Joseph, 2020), virtues that refuse social demands to 'be quiet' and that use imaginative futures grounded in the strengths and brilliance of Black learners write the histories of those futures as they are being made. We are making these futures. Afrofuturism takes its form as science fiction and fantasy rooted in Black history. The writers N. K. Jemisin and Nnedi Okofor are often considered Afrofurists. So is American singer, rapper and actress Janelle Monáe. So are writer Alan Moore, artist Dave Gibbons and colorist John Higgins, the British creative team that produced the original *Watchmen* graphic book maxiseries.

Nathan Alexander, professor of mathematics education at Moorehouse College, urges us to seriously look at the possibilities for mathematics outside of what we know. He asks us to imagine, in good Afrofuturist style, "What would math be like in our (fictional) future where mathematics had developed for a long time rooted in Black life?" He himself did a kind of mind experiment inspired by Derrick Bell, a scholar who had told a story of a special island only reached by Black people where there is no racism and no history of oppression. Alexander imagined what it is like for his invented characters, Deja and her friend, to fly from planet to planet using equations and mathematical modeling, in a universe where there is never a question about whether or not certain students are achieving (Alexander, 2019) … because, after all, if that is the question, then we are maintaining the other side of the question, the implicit assumption that Black students are very possibly not going to perform well in mathematics. So: equity is no longer the theme. What is, then? We think *beyond*. What does the mathematics classroom look like? What are people doing? There is probably no need to fixate on social justice mathematics, for example. What *do* people do in these classrooms? The mathematics is created by the brilliant Black children and young adults in their math studio classrooms, and it is spread all over the universe along with the Black Joy that inspired it.

Reconstitution

... involves breaking down and reorganizing; disorienting and reorienting; it means opening up and expanding frames of theory and politics", writes Professor Zamalin (2022). By undoing in fiction what is socially familiar and naturalized in practice and in history – whether unconscious racist common sense or the sense of political and legal stasis – Afrofuturism opens up a vantage for a utopian reconsideration of the (racially unequal) status quo. Throughout its long and complicated history, the utopia Afrofuturism posits isn't characterized by a romantic notion of progress or tranquility. It directly insists on describing and detailing a future that this present is going to make. Keeping in mind the ways that racism and white supremacy consistently undermine Black freedom dreams and antiracist horizons, Afrofuturism imagines utopia as a conscious refusal of the tensions. Okay, we can't forget the contradictions and tragedies that make contemporary life 'the reality'. Yet we can still believe in this optimistic vision, and *make the history* that creates it. Beyond refreshing our resolve to imagine the impossible and work to bring it into existence, reconstitution as a virtue uses the long history of Afrofuturism to place our students in the position of being *the* ones. Our students are the ones who were first identified later by historians as triggering this future utopia. They created mathematics that made the impossible possible. This antiracist future became real because of our students. We helped to make it happen.

Refraction of Imagination, Technology & Liberation

... are together the processes that we commit to by pursuing the Afrofuturist virtues. We and our students together take imagination, technology and liberation through ourselves as if we are prisms, and they come out of us in beautiful algorithms, social concepts, and as yet inconceivable tools for living, much like light is refracted into a rainbow of colors by a physical prism. Our students mix together mathematical relationships, objects, tools, arguments, questions, and paradoxes, and *woosh*, new mathematics is invented. The Afrofuturist virtues enable this generative focus for re-imagining mathematics co-production because of their diverse, living and lively practices and applications (Eseonu & Duggan, 2021). Afrofuturist mathematics "combines elements of science fiction, historical fiction, speculative fiction, fantasy, Afrocentricity, and magic realism with non-Western beliefs" (Womack, 2013) to reimagine, re-constitute, re-present and re-invent meaningful black subjectivities.

World-Building Retrodiction

"Retrodiction" is the act of making a prediction about an imaginary past for a possible future. In general, archaeologists or geologists might use information or ideas from the present to infer or explain a past event or state of affairs. Our commitment to living the virtue of Retrodiction means that we infer from the as-yet fictional future the "past" that created it, and we *make that past happen now*, so that the fictional future inspires us and propels us into some sort of future that shares many characteristics of our fictional future. The lists in this book that feel the most far-fetched to you right now may need to become the lists to which you most strongly commit, in order to achieve these futures.

Collaborative Appropriation with Community Elders

Afrofuturism provides an ecosystem of ideas, practices, critical philosophy and liberating strategies that challenge the imaginative underpinnings of racial oppression. It provides an archive of imaginative work that we all can interact with as inspiration for creative classroom mathematics. Afrofuturism provides a coherent framework of philosophical orientations and transmedia practices for re-imagining the institutional context of school mathematics. It is especially useful for freshly re-orienting ourselves to elite, Eurocentric institutions and discourses (See the discussions of Ethnomathematics and Critical Mathematics Education in "Levels of Democratic Mathematics" in Chapter 7). Its urgent call is to re-imagine our world as if it had never experience the tragedy of the transatlantic slave trade. However, what kinds of engagements by whom are appropriate? There is a way that Afrofuturist mathematics could fall prey to coloniality similar to those unaware applications of Ethnomathematics that are metaphorically raping non-Western cultures for intellectual resources. During colonialism, Western powers mined colonies for metals, spices, and human labor. We do not want to "mine" Afrofuturism for its fresh and creative potential. How do we manage this? By consistently challenging ourselves to question our motives. One ideal virtue is *Collaborative Appropriation with Community Elders* who initiate our Afrofuturist-inspired classroom activity.

Further Reading

Alexander, N. (2019). Daija's awakening: Critical race theory and Afrofuturism in mathematics education. In J. Davis & C. C. Jett (Eds.), *Critical race theory in mathematics education* (pp. 56–74). Routledge.

Eseonu, T., & Duggan, J. (2021). Negotiating cultural appropriation while reimagining co-production via Afrofuturism. *Qualitative Research Journal, 22*(1), 96–107.

Joseph, C. (2020, July 29). What black joy means—and why it's more important than ever. *British Vogue*. https://www.vogue.co.uk/arts-and-lifestyle/article/what-is-black-joy

Worsley, T., Roby, R. (2021). Fostering spaces for black joy in STEM-rich making and beyond. *Journal of Effective Teaching in Higher Education, 4*(2), 118–134.

Zamalin, A. (2022). Afrofuturism as reconstitution. *Afrofuturism and the Law, 9*(1). https://cal.library.utoronto.ca/index.php/cal/article/view/38263

Shopping Lists

What Every Mathematics Classroom Never Knew It Needed

- Paper Clips
- String
- Cabinet of Curiosities
- MathTour Library
- Social Media Identity
- Community Advisory Board

Items on this list went through a careful vetting process. Only the most essential survived the scrutiny of the author and his students over the course of 40 years of teaching and collaborating with teachers, and friends.

Paper Clips

Lots of them. First of all, they're handy for keeping slips of paper together in a way that is easily changed and adjusted, added to, culled, etc. But more importantly, they are perfect for making *Spinners*. Forget dice of different-numbered sides, forget pre-made spinners that you never have around. Just grab a pen, a scrap of paper, and draw a shape. Put a dot somewhere inside, and draw lines radiating out from that dot. Use the pen, through the clip, on the dot, to hold the clip in place, and spin. You can create a spinner with any number of outcomes, spinners with unequal outcomes that have some bigger sections or smaller sections, spinners with some sections where the player of a game decides the outcome, and so on. Students can design games and see how a different set of possible moves changes the game to be better or worse, fairer or rigged, that compensate for unequal starting situations, and more. There is no end to clip-spinner creativity.

String

Lots of it, preferably different colors, so that the color can be used to represent something different. First of all, string is reusable for tying things together and keeping them in place. It can be used to hang things from the ceiling with one of those paper clips hooked onto those frames of the ubiquitous drop ceilings found in so many public buildings and schools. More importantly, a group of students can be challenged to together create 2-, 3- or models of n-dimensional shapes. They pull the string taut at the vertices with hands and other body parts. String can model 2- and 3-dimensional graphs and networks, highlighting the paths rather than the nodes, skeletons of 3-dimensional shapes, or concept map connections for mathematical topics. Any length can be a non-standard unit of distance measure that easily conforms to a curved perimeter. String can be used to mathematize knots or braids, to create a pendulum or to study the physics of sound.

Cabinet of Curiosities

Connecting to the theme of Resonance and Wonder (See "Storytelling Techniques for Math Teachers", Chapter 8), the origin of this museum curating strategy is found in the popularity of cabinets of curiosity during the Italian Renaissance. Cabinets of curiosities were just the thing for an aristocrat who didn't want their guests to think their party was boring. Ranging in size, they might be any sort of dedicated piece of furniture with multiple drawers. Some would fill a room. These "cabinets" held unique and special objects acquired through long journeys to faraway lands. Each object offered an opportunity to tell a story about an epic adventure or, more often, to invent one. The wealthy liked to define their personalities through the possession of glamorous objects as tangible tokens of their intelligence, erudition, wealth, and taste. They weren't always built on truth – artifacts and preserved animals and plants were sometimes purchased at the markets in Florence and Rome rather than collected directly by the owner of the cabinet. Some, like the famous mermaids made by stitching together the torso of a monkey to the tail of a fish were wholly fantastical fabrications closer to art than natural evidence. Truth wasn't the point. In spirit, the cabinets were not meant to be scientific – they were a place of the imagination in which those who could afford to do so, constructed their own personal versions of the world (See "Afrofuturist Virtues). "Standing at the center of this mini-universe and pointing at the objects to disclose their deepest secrets, collectors felt a sense of ease and mastery over a world that most often appeared too big, too confusing, and too inhospitable", writes art historian Aloi (2018).

Every math classroom should have its own cabinet of curiosities. A closed cabinet, a box, a drawer, anything can be the "cabinet". Teachers can collect artifacts from their own mathematics journeys, their childhoods, photos from their travels around the world that depict math themes or places of special interest to mathematics, and bring them out at special times of revelation. "Ooh, I actually took a picture of myself with my son at the Parthenon in Athens, Greece, a site of "Golden Rectangles". "Here I am eating ice cream at the *Gauss Turm* near Göttingen, Germany". "I actually used this circular slide rule in Chemistry class in high school! We all did! It said 'Chemistry is *chool*' on the sleeve that I lost since then!" "I made these 'Napier's Bones' in a college class I took way back when". "This wooden box of 'math baggage' was made by a student about 7 years ago – inside it has anxieties, pictures of the textbook pages she wanted to tear her hair out over, and so on – she left it with me, she said, because she left her baggage behind and had a new and happy outlook on math!".

Math Tour Library

No math classroom is complete without a library of math tours. These are guides for walking, bike riding, journeys on public transportation, or for family outings, with points of interest and accompanying commentary that provokes thinking about mathematics at each stopping point. A teacher can create some of these to start. Ideally, though, students over the years will add to the library. A good model is found in those "Little Free Libraries" that are in some neighborhoods, where you can leave a book and take another. Maybe students might be required to contribute one of their own guides in order to borrow one from the library. Or perhaps small groups of students create a math tour for each unit of the year, specifically finding ways to incorporate the skills and concepts for the unit into their math tour. There might be tours of the classroom, tours of the school, tours of the town or city… students might bring back math tour guides if they travel during vacations, for far-away locations. Tours might have historical themes, emotional math themes, environmental justice themes – the key is that the thinking at each site should involve mathematics and its uses. Ideally, the tours incorporate all sorts of ways of approaching mathematics, including problem-solving, mathematical thinking with patterns and conjectures, and problem-posing (See "Problem Solving and Posing Lists").

Social Media Identity

A group social media identity makes all sorts of outreach possible. Students can post classroom news on a social media site, which requires them to articulate what they are learning and what exciting math fashion is popular in the classroom at the moment. They can follow mathematicians on key sites (See "First Aid Supplies for the Math Teacher's Body & Spirit" for some suggestions). The class can host a "Problem of the Week" or "Challenge of the Month" through which they act as mathematics ambassadors, attracting folks from around the world to dialogue on good problems, clever solutions, and interesting follow-up questions. For cyber-safety reasons, the group identity allows the class to gain from social media without needing to expose the real-life identity of any individual student. They might exchange problems, posed or solved, with other classes around the world, with elders in far-away communities, and more.

Community Advisory Board

Now is always the time to start or add to your Community Advisory Board. This is a collection of folks in the surrounding region who agree to be available for questions, advice, or even for classroom visits – mathematicians, engineers and scientists; professionals like architects, dentists, salon hair colorists, carpenters, gardeners, zookeepers and urban planners; artists and government policy workers; activists and school neighbors. Advisory board members agree to take a phone call or answer an email message asking for tips, advice, or information. Membership might involve participation on a panel of evaluators for class projects. It's always good to be clear that you won't abuse their membership – maybe something like, no more than one request per month, unless we have your permission to follow-up, or "you can always refuse an invitation to visit, although if so we would love it if you might know someone else who is available".

Further Reading

Association for Women in Mathematics. (n.d.). *Mentor network*. https://awm-math.org/programs/mentor-network/

Benjamin Banneker Association. (n.d.). *Mathematics education for all children, especially African-American children*. https://www.benjaminbannekerassociation.org/

Borden, C. (n.d.). *Implementing effective youth mentoring relationships for high school students*. U.S. Department of Education. https://www2.ed.gov/programs/slcp/finalimplem.pdf

Franklin Institute. (n.d.). *The math tour*. https://www.fi.edu/en/math-tour

National Museum of Mathematics. (n.d.). *Mo Math*. https://momath.org/
TODOS: Mathematics for All. (n.d.). *Excellence and equity in mathematics*. https://www.todos-math.org/

First Aid Supplies for the Math Teacher's Body & Spirit

- First Aid Kit
- Ben Orlin's *Math with Bad Drawings*
- Twitter account following Cliff Pickover
- A free substack subscription to Dan Meyer's MathWorlds
- YouTube account following Eddie Woo, Kyne Santos, Vi Hart, Numberphile, Mathologer, & Chris Staecker
- A *Freakonomics* network membership

The information contained in this book, including but not limited to, text, graphics, images and other material, are for informational and entertainment purposes only. No material on this list is intended as a substitute for professional medical advice, diagnosis, or treatment, spiritual guidance, or career coaching.

First Aid Kit

The American Red Cross recommends keeping a first aid kit in your home and car. They further recommend that you carry a first aid kit with you or know where you can find one at all times, and that you find out the location of the first aid kit where you work. They would surely say there should be one in your classroom. If you cannot find one there, here are the recommended items:

- 2 absorbent compress dressings (5 × 9 inches)
- 25 adhesive bandages (assorted sizes), also found within our Family First Aid Kit
- 1 adhesive cloth tape (10 yards × 1 inch)
- 5 antibiotic ointment packets (approximately 1 gram)
- 5 antiseptic wipe packets
- 2 packets of aspirin (81 mg each)
- 1 emergency blanket
- 1 breathing barrier (with one-way valve)
- 1 instant cold compress
- 2 pair of non-latex gloves (size: large)
- 2 hydrocortisone ointment packets (approximately 1 gram each)
- 1 3 in. gauze roll (roller) bandage
- 1 roller bandage (4 inches wide)
- 5 3 in. × 3 in. sterile gauze pads
- 5 sterile gauze pads (4 × 4 inches)

– Oral thermometer (non-mercury/nonglass)
– 2 triangular bandages
– Tweezers
– Emergency First Aid instructions

Whether you buy a first aid kit or put one together, make sure it has all the items you may need: Include any personal items such as medications and emergency phone numbers or other items your health-care provider m ay suggest in your own, personal, first-aid kit, apart from one that is accessible to students; check kits regularly; and routinely check expiration dates – replace any used or out-of-date contents.

Ben Orlin's Math with Bad Drawings

Ben Orlin's (2018) *Math with Bad Drawings: Illuminating the ideas that shape our reality* (Black Dog & Levinthal Publishers) is hilarious and insightful at the same time. Sure his boog is great. But the book is portable and fits into a purse or backpack. And it's more extensive and polished than the blog, being a revision of the blog items. Whenever you're feeling a little down or disappointed in something, open this book. Math for the mind! Uplifting, indeed.

Cliff Pickover Books and Social Media

Cliff Pickover is a mathematician and author of a large number of popular mathematics books that are accessible to non-mathematicians. He posts on *X* (formerly Twitter) more than once a day, and it's always entertaining. It only take a minute to follow him. I personally avoid the comments, which I find less than tantalizing. But his posts are curious, often funny, and sometimes provoke some distracting math thinking.

Dan Meyer Social Media

Dan Meyer is a popular speaker at mathematics education conferences because he is *always* inspiring. Dan is a mathematics teacher known for the idea of "be less helpful" (See "Promoting a Thinking Classroom"). Since he gave the lecture from which that fame emerged, he has created a fantastic blog, hosts an interesting podcast with other mathematics teachers, and basically has consistently wonderful ideas for helping students love mathematics and engage in mathematical inquiry.

There are tons of clever Vloggers (Video bloggers) out there, who focus on mathematics. These five met the serious and careful vetting of the non-existent advisory board for this book: *Eddie Woo, Kyne Santos, Vi Hart, Numberphile, Mathologer, & Chris Staecker*. Find them on YouTube and watch their videos.

They are astoundingly clever, usually hysterically funny, and supply unending ideas for stories you can tell in your class. I often show the videos themselves *in* my class, or recommend them to students and families. Eddie Woo, a Australian mathematics teacher, started our sharing some of his own lessons on social media, and has taken on the task of helping everyone find joy in the teaching and learning of mathematics. Kyne Santos is a TikTok and Instagram drag queen star bringing math and drag and queer culture all together to a world-wide audience. Vi Hart is a U.S.-American "mathemusician", and defies description. Numberphile is a project of British journalist Brady Haran, and is sponsored by several British scientific foundations, interviewing in entertaining ways a wide range of diverse mathematicians about easily accessible mathematics topics; the mathematicians are exemplary models of mathematical thinking and mathematical curiosity. The "Mathologer" is Bukard Polster, a real-life mathematics professor at Monash University in Melbourne, Australia; he describes his frequent YouTube videos as "really accessible explanations of hard and beautiful math(s)". Chris Staecker is a U.S.-American professor of mathematics who has two kinds of videos that he posts: in some, he reviews the depiction of mathematics that his followers find in media, mostly television and film; in others, he reviews no obscure mathematical tools and devices that were once readily available in the world and are now their own curious openings into highly specific mathematical topics (See "Cabinet of Curiosities" in "What Every Math Classroom Never Knew it Needed").

A Freakonomics Membership

There are books by Steven Levitt and Stephen Dubner with *Freakonomics* in their titles. They are nice. However, the *Freakonomics Radio* podcast is fantastic for "recovery-mode" or commuter listening. Co-author Stephen J. Dubner uncovers the hidden side of everything. Why is it safer to fly in an airplane than drive a car? How do we decide whom to marry? Why is the media so full of bad news? Also: things you never knew you wanted to know about wolves, bananas, pollution, search engines, and the quirks of human behavior. Data and its analysis, mathematical inferences, deduction, conjectures, and the paradoxes and complexities of the differences between intuition and mathematical results is made into sublime entertainment in this podcast.

Further Reading

Freakonomics Radio. (n.d.). https://freakonomics.com/series/freakonomics-radio/

Numberphile. (n.d.). https://www.numberphile.com/

OnlineKyne. (n.d.). *Instagram feed.* https://www.instagram.com/onlinekyne/?hl=en
Orlin, B. (n.d.). *Math with Bad Drawings: Lover of math. Bad at drawing*. Blog. https://mathwithbaddrawings.com/author/mathwithbaddrawings/
WooTube. (n.d.). *Website of Eddie Woo.* https://misterwootube.com/

A Short List of Mathematical Games You Might Have Overlooked

- Ravensburger Labyrinth Family Board Game
- Mental Jotto
- Looney Labs Fluxx
- Hasbro Clue
- Set Enterprises Set Game
- Perplext Pack-O-Games

These games are not "math games" – they don't require mathematical calculations or solutions to mathematical challenges. They are instead 'metaphorical games' for mathematics. Playing these games requires mathematical thinking. The could be templates for actual mathematical games, if you or your students design games with the same components, but replace the cards, pieces, board locations, and so on, with mathematical facts, questions, algorithms, patterns, topics, paradoxes, etc. The play of these games creates the need to link across relationships and to conjecture like a mathematician (See "Playbook Lists").

Ravensburger Labyrinth Family Board Game

In this game, players plan paths to treasures in a labyrinth. In order to create a possible path, they slide pieces of the board that adjust entire rows or columns of the board itself. One cannot play this game without developing a fine sense of two-dimensional Cartesian coordinate systems, and curved pathways within them. Strategy also involves thwarting others' needed pathways, and the understanding of a wrap-around coordinate system similar to video games.

Mental Jotto

This game is like playing Mindscape in your head, and a bit like the recently popular online *Wordle* game. Two or more players think of a secret 5-letter word before the game begins. They then take turns guessing a five-letter word for one of their opponents. For two players, they alternate. For three or more, the simplest version is to query the person to your right, with play proceeding counter-clockwise; although a more competitive and memory-taxing version allows a player to query any other player. The other player then honestly says how many of the letters are included in their secret word – unlike the *Wordle* game, one does not know about correct placement, only correct inclusion. First person to guess the other player's word wins. In a multi-player game, the first player whose word is guessed is out; last player to have their word guessed wins. Strategy discussion is key to making this mathematical. Testing out strategies is a critical feature of the mathematization for this game. One can also explore the question: why *five* letters? Collect data from how games with 2–4 and 5–7 letters unfold, and establish conjectures to test (See "John Mason, Kaye Stacey, Leone Burton's *Thinking Mathematically*"). I usually introduce the game by recording guesses and the corresponding number of correct letters on the board or on paper, but the game is far more interesting when played only in one's head.

Looney Labs Fluxx

The basic Fluxx game is the ultimate metaphor for mathematics as a "game" with arbitrary objects and rules for how to operate upon them. It is a deck of cards with items called "Keepers", Rule cards and other cards that create transformations upon the rules. The game starts with a basic rule: Take one card and play one card. From then on, the game rules change as part of the play of the game, including the goal of the game. Looney Labs produces numerous versions with variously themed objects. Once you've played this game, it is a small step further to imagine creating your own version of it. I use index cards, but any small blank material that you can write and draw on to make a deck of "cards" will work. Create the deck using mathematical objects from your classroom as the "Keepers", and the functions, applications and other uses of mathematics in class as the rules for how they can be combined or transformed. Students can design their own versions as well! One fun way to create a stack of Keepers is to have students continue to add new cards (shapes, functions, operations, problem types, graphs and other images, etc.) every week or month to the deck that a small group will use to play their Math Fluxx throughout the year.

Clue & Set

Hasbro Clue and the *Set Enterprises Set Game* can work like Fluxx – as the template for a mathematics game where each component is replaced by mathematical resources from the current curriculum. The structure of Clue uses a basic "board" with "locations" – the places might be strands of mathematics (Geometry-Algebra-Word Problems-Graphing Techniques), realms of numbers (Counting Numbers-Rationals-Irrationals – Transcendentals or Whole

Numbers-Fractions-Decimals-Negative Numbers), Spaces (2D-3D-Sphere-Torus), and the players move to a location on their turn. There is a mystery to be solved with several attributes, each of which have several options, and the goal is to figure out the secret combination: which degree polynomial with what coefficient of x, and what constant? What speed of which object on which slope for how far? What shape with which area and which perimeter? Anything is possible. Set is often played because it feels math-y, but again, the treasure is in its use as a template: A teacher and/or their students can create their own deck of "mathematical thing" cards from the curriculum so far experienced this year. Just like set, a small group plays by dealing face-up, say three rows of four cards. If a player recognizes a "set", they say "set" and explain the three cards that make the set. The fundamental rule for this, just like in the original game, is, "Whatever any two have in common, all have in common". In a mathematics version, this is simply done by students' creativity. They must convince the others that the rule works for their set (See "Convince Yourself, Convince a Friend, Convince a Skeptic" in the discussion of "John Mason, Kaye Stacey & Leone Burton's *Thinking Mathematically*", Chapter 8). Of course, one of their opponents can challenge the set, if they can see something that is not in common for the entire set; but they, too, must convince the others of what they believe.

Perplext Pack-O-Games

Perplext Pack-O-Games are popular among mathematics teachers as pastimes and as occasional gifts. Their clever gimmick is that they come as a pack of cards the size and shape of sticks of gum, in tiny chewing gum packages. Game play makes use of the cards as a playing board under construction throughout the game, and as playing components. Hopefully the company will continue to come up with new ideas. While some are mathematical in and of themselves, each is a template for students to invent their own games, and they function well as examples of everyday small materials. Nothing special is needed to design a game. The mathematics comes in the revision of multiple game drafts – is this a good game yet? What would make it just the right level of challenge? Are there variations once players get bored with the original? Can it be more complex or simpler? What if you replace some components with others (See "Stephen I. Brown & Marion Walter's *Art of Problem Posing*" in the "Problem-Solving & Posing Lists" in Chapter 8)? Students can test out each other's' games and provide feedback for further design revisions.

Further Reading

Orlin, B. (2022). *Math games with bad drawings. 75 ¼ simple, challenging, go-anywhere games – and why they matter.* Black Dog & Leventhal.

Wells, D. (2012). *Games and mathematics: Subtle connections.* Cambridge University Press.

CHAPTER 10

List of Last Thoughts

- Start Over
- Build-a-Book
- To-Do Lists

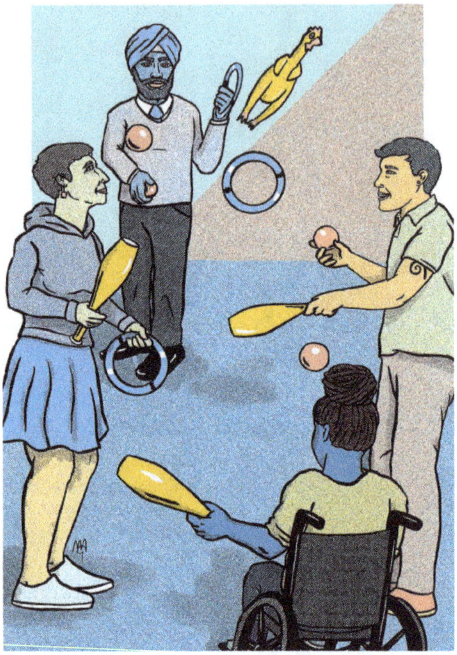

As in the option to "Choose Your Own Adventure" in the list of "How to Read this Book", what you can do at the end of the book really opens up a new list:

- *Start Over* – Go back to that first list of how to read the book and try another way to use it.
- *Build-a-Book* – Just like those shopping mall stores where you can make your own Bears out of your own choice of provided materials, and a little like that crazy idea of making a list poem out of randomly found sentences in the book, but with a lot more structure and a huge ingredient of your own wisdom: Take pieces from different sections of this book and organize them into *your own lists*: reorganize the ideas and the suggested strategies to fit your own approach to teaching mathematics. This will be similar to the idea of categorizing problems by attributes in "Stephen I. Brown & Marion Walter's

© PETER APPELBAUM, 2024 | DOI:10.1163/9789004688469_011

Art of Problem Posing", where students would first organize the problems one way, and then another. I did the first organization in this book. Now you can reorganize ideas and concepts according to very different attributes!
- *To-Do Lists* – Make three To-Do lists. What are you going to do:
  - Today
  - In the next two weeks
  - In the next year

... to follow up on the goals you developed while using this book.

References

Aloi, G. (2018, August 28). *Cabinets of curiosities and the origin of collecting*. Sotheby's Institute of Art. https://www.sothebysinstitute.com/news-and-events/news/cabinets-of-curiosities-and-the-origin-of-collecting

Appelbaum, P. (2000). Eight critical points for mathematics. In D. Weil (Ed.), *Perspectives in critical thinking: Essays by teachers in theory and practice* (pp. 41–55).

Appelbaum, P. (2018). The curricular geometries of *SAMBA* schools: Fractal dimensions, surface, depth, and recursion. In S. Gerofsky (Ed.), *Contemporary environmental and mathematical education modelling using new geometric approaches* (pp. 67–80). Palgrave.

Arendt, H. (1958). *The human condition.* The University of Chicago Press.

Baldinger, E. M., Louie, N., & the Algebra Teaching Study and Mathematics Assessment Project. (2016). *The TRU Math conversation guide: A tool for teacher learning and growth.* https://truframework.org/

Bishop, A. J. (1988). *Mathematical enculturation: A cultural perspective on mathematics education* (pp. 100–103).

DePiper, J. N., Louie, J., Nikula, J., Buffington, P., Tierney-Fife, P., & Driscoll, M. (2021). Promoting teacher self-efficacy for supporting English learners in mathematics: Effects of the visual access to mathematics professional development. *ZDM-Mathematics Education, 53*, 489–502.

Dery, M. (1994). Black to the future: Interviews with Samuel R. Delaney, Greg Tate, and Tricia Rose. In M. Dery (Ed.), *Flame wars: The discourse of cyberculture.* Duke University Press. https://doi.org/10.1215/9780822396765-010

Du Bois, W. E. B. (1900). *African American photographs assembled for 1900 Paris Exposition.* Library of Congress Prints and Photographs Online Collections. https://www.loc.gov/pictures/collection/anedub/dubois.html

Dweck, C. (2007). *Mindset: The new psychology of success.* Random House.

Frey, J. (2001, November 18). What is virtue and why does it Matter? *Flypaper.* https://fordhaminstitute.org/national/commentary/what-virtue-and-why-does-it-matter

Ginsburg, H., Jacobs, S., & Lopez, L. S. (1998). *The teacher's guide to flexible interviewing in the classroom: Learning what children know about math.* Allyn & Bacon.

González, N., Moll, L. & Amanti, C. (Eds.). (2005). *Funds of knowledge: Theorizing practices in households, communities, and classrooms.* Erlbaum.

Gordon, M. (2007). Living the questions: Rilke's challenge to our quest for certainty. *Educational Theory, 57*(1), 37–51.

Herek, S. (1989). *Bill and Ted's excellent adventure.* Film. Screenplay by Solomon and Chris Matheson (Ed.). Orion Pictures.

Keitel, C., Klotzmann, E., & Skovsmose, O. (1993). Beyond the tunnel vision: Analyzing the relationship between mathematics, society and technology. In C. Keitel & K. Ruthven (Eds.), *Learning from computers: Mathematics education and technology* (pp. 243–279). Springer.

Kirshner, D. (2002). Untangling teachers' diverse aspirations for student learning: A crossdisciplinary strategy for relating psychological theory to pedagogical practice. *Journal for Research in Mathematics Education, 33*(1), 46–58.

Knott, M. L. (2015). *Unlearning with Hannah Arendt.* Other Press.

Ladson-Billings, G. (1994). *Dreamkeepers: Successful teachers of African American children.* Jossey-Bass.

Ladson-Billings, G. (1995). Making mathematics meaningful in multicultural contexts. In W. Secada, E. Fennema, & L. Byrd (Eds.), *New directions for equity in school mathematics* (pp. 126–145). Cambridge University Press.

Meaney, T. (2018). *Ensuring social justice: Digitalizing mathematics learning opportunities with a focus on multilingual children.* KINDknow – Kindergarten Knowledge Centre for Systemic Research, https://www.youtube.com/watch?v=6x8Jj-R95Oo

Ohanian, S. (1992). *Garbage pizza, patchwork quilts, and math magic stories about teachers who Love to Teach and Children who Love to Learn.* W.H. Freeman.

Rilke, R. M. (1984). *Letters to a young poet* (S. Mitchell, Trans.). Random House.

Sontag, S. (1966). *Against interpretation and other essays.* Farrar, Strauss, & Giroux.

Verhulst, F. (2012). Mathematics is the art of giving the same name to different things: An interview with Henri Poincaré. *Nieuw Archief Voor Wiskunde, Serie 5, 13*(3), 154–158.

Wells, I. (1894). *A red record: Tabulated statistics and alleged causes of lynchings in the United States, 1892-1893-1894.* Donohue & Henneberry. https://www.gutenberg.org/files/14977/14977-h/14977-h.htm

Womack, Y. (2013). *Afrofuturism: The world of black Sci-Fi and fantasy culture.* Chicago Review Press.

Womack, Y. (2018, November 27). What is Afrofuturism? Mixing sci-fi, magic and non-Western beliefs, Afrofuturism frees individuals. *IAI News.* https://iai.tv/articles/what-is-afrofuturism-auid-1177

Index of Mathematical Topics

Index of Names

www.ingramcontent.com/pod-product-compliance
Lightning Source LLC
Chambersburg PA
CBHW071225230926
51849CB00001B/32
9789004688445